# Starting Electronics

# Starting
# Electronics

## Keith Brindley

AMSTERDAM • BOSTON • HEIDELBERG • LONDON • NEW YORK •OXFORD
PARIS • SAN DIEGO • SAN FRANCISCO • SINGAPORE • SYDNEY • TOKYO
Newnes is an imprint of Elsevier

Newnes
An imprint of Elsevier
Linacre House, Jordan Hill, Oxford OX2 8DP
200 Wheeler Road, Burlington, MA 01803

First published 1994
Second edition 1999
Third edition 2005
Copyright © Keith Brindley 1994, 1999, 2005. All rights reserved.

**British Library Cataloguing in Publication Data**
A catalogue record for this book is available from the British
Library

ISBN 07506 63863

Typeset and produced by Co-publications, Loughborough

# Contents

# Preface

This book originated as a collection of feature articles, previously published as magazine articles. They were chosen for publication in book form not only because they were so popular with readers in their original magazine appearances but also because they are so relevant in the field of introductory electronics — a subject area in which it is evermore difficult to find information of a technical, knowledgeable, yet understandable nature. This book is exactly that. Since its original publication, I have added significant new material to make sure it is all still highly relevant and up-to-date.

I hope you will agree that the practical nature of the book lends itself to a self-learning experience that readers can follow in a logical, and easily manageable manner.

# 1 The very first steps

Most people look at an electronic circuit diagram, or a printed circuit board, and have no idea what they are. One component on the board, and one little squiggle on the diagram, looks much as another. For them, electronics is a black art, practised by weird techies, spouting untranslatable jargon and abbreviations which make absolutely no sense whatsoever to the rest of us in the real world.

But this needn't be! Electronics is not a black art — it's just a science. And like any other science — chemistry, physics or whatever — you only need to know the rules to know what's happening. What's more, if you know the rules you're set to gain an awful lot of enjoyment from it because, unlike many

# Starting electronics

sciences, electronics is a practical one; more so than just about any other science. The scientific rules which electronics is built on are few and far between, and many of them don't even have to be considered when we deal in components and circuits. Most of the things you need to know about components and the ways they can be connected together are simply mechanical and don't involve complicated formulae or theories at all.

That's why electronics is a hobby which can be immensely rewarding. Knowing just a few things, you can set about building your own circuits. You can understand how many modern electronic appliances work, and you can even design you own. I'm not saying you'll be an electronics whizz-kid, of course — it really does take a lot of studying, probably a university degree, and at least several years' experience, to be that — but what I am saying is that there's lots you can do with just a little practical knowledge. That's what this book is all about — starting electronics. The rest is up to you.

## What you need

Obviously, you'll need some basic tools and equipment. Just exactly what these are and how much they cost depends primarily on quality. But some of these tools, as you'll see in the next few pages, are pretty reasonably priced, and well worth having. Other expensive tools and equipment which the professionals often have can usually be substituted with tools or equipment costing only a fraction of the price. So, as you'll see, electronics is not an expensive hobby. Indeed, its

# The very first steps

potential reward in terms of enjoyment and satisfaction can often be significantly greater than its cost.

In this first chapter I'll give you a rundown of all the important tools and equipment: the ones you really do need. There's also some rough guidelines to their cost, so you'll know what you'll have to pay. Tools and equipment we describe here, however, are the most useful ones you'll ever need and chances are you'll be using them as long as you're interested in electronics. For example, I'm still using the side-cutters I got over twenty years ago. That's got to be good value for money.

## Tools of the trade

Talking of cutters, that's the first tool you need. There are many types of cutters but the most useful sorts are side-cutters. Generally speaking, buy a small pair — the larger

Photo 1.1    Side-cutters like these are essential tools – buy the best you can afford

3

ones are OK for cutting thick wires but not for much else. In electronics most wires you want to cut are thin so, for most things, the smaller the cutters the better.

You can expect to pay from £4 up to about £50 or so for a good quality pair, so look around and decide how much you want to spend.

**Hint:**

If you buy a small pair of side-cutters (as recommended) don't use them for cutting thick wires, or you'll find they won't last very long, and you'll have wasted your money.

You can use side-cutters for stripping insulation from wires, too, if you're careful. But a proper wire stripping tool makes the job much easier, and you won't cut through the wires underneath the insulation (which side-cutters are prone to do) either. There are many different types of wire strippers ranging in price from around £3 to (wait for it!) over £100. Of course, if you don't mind paying large dentist's bills you can always use your teeth — but certainly don't say I said so. You didn't hear that from me, did you?

A small pair of pliers is useful for lightly gripping components and the like. Flat-nosed or, better still, snipe-nosed varieties are preferable, costing between about £4 to £50 or so. Like

**Photo 1.2   Snipe-nosed pliers – ideal for electronics work and another essential tool**

side-cutters, however, these are not meant for heavy-duty engineering work. Look after them and they'll look after you.

The last essential tool we're going to look at now is a soldering iron. Soldering is the process used to connect electronic

**Photo 1.3   Low wattage soldering iron intended for electronics**

components together, in a good permanent joint. Although we don't actually look at soldering at all here, a good soldering iron is still a useful tool to have. Soldering irons range in price from about $4 to (gulp!) about $150, but — fortunately — the price doesn't necessarily reflect how useful they are in electronics. This is because irons used in electronics generally should be of pretty low power rating, because too much heat doesn't make any better a joint where tiny electronic components are concerned, and you run the risk of damaging the components, too. Power rating will usually be specified on the iron or its packing and a useful iron will be around 15 watts (which may be marked 15 W).

It's possible to get soldering irons rated up to and over 100 watts, but these are of no use to you — stick with an iron with a power rating of no more than 25 watts. Because of this low power need, you should be able to pick up a good iron for around a tenner.

These are all the tools we are going to look at in this chapter (I've already spent lots of your money — you'll need a breather to recover), but later on I'll be giving details of other tools and equipment which will be extremely useful to you.

## Ideas about electricity

Electricity is a funny thing. Even though we know how to use it, how to make it do work for us, to amplify, to switch, to control, to create light or heat (you'll find out about all of these aspects of electricity over the coming chapters) we can still only guess at what it is. It's actually impossible to see electricity: we only see what it does! Sure, everyone knows that electricity is a flow of electrons, but what are electrons? Have you ever seen one? Do you know what they look like?

The truth of the matter is that we can only hypothesise about electricity. Fortunately, the hypothesis can be seen to stand in all of the aspects of electricity and electronics we are likely to look at, so to all intents and purposes the hypothesis we have is absolute. This means we can build up ideas about electricity and be fairly sure they are correct.

Right then, let's move on to the first idea: that electricity is a flow of electrons. To put it another way, any flow of electrons is electricity. If we can measure the electricity, we must therefore be able to say how many electrons were in the flow. Think of an analogy, say, the flow of water through a pipe (Figure 1.1). The water has an evenly distributed number of foreign bodies in it. Let's say there are ten foreign bodies (all right then, ten specks of dust) in every cm³ of water.

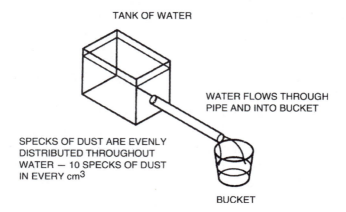

TANK OF WATER

WATER FLOWS THROUGH
PIPE AND INTO BUCKET

SPECKS OF DUST ARE EVENLY
DISTRIBUTED THROUGHOUT
WATER — 10 SPECKS OF DUST
IN EVERY cm$^3$

BUCKET

**Figure 1.1  Water flowing in a pipe is like electricity in a wire**

# Starting electronics

Now, if 1 litre of water pours out of the end of the pipe into the bucket shown in Figure 1.1, we can calculate the number of specks of dust which have flowed through the pipe. There's, as near as dammit, 1000 cm³ of water in a litre, so:

$$10 \times 1000 = 10{,}000$$

water-borne specks of dust must have flowed through the pipe.

Alternatively, by knowing the number of specks of dust which have flowed through the pipe, we can calculate the volume of water. If, for example, 25,000 specks of dust have flowed, then 2.5 litres of water will be in the bucket.

## Charge

It's the same with electricity, except that we measure an amount of electricity not as a volume in litres, but as a charge in coulombs (pronounced koo-looms). The foreign bodies which make up the charge are, of course, electrons.

There's a definite relationship between electrons and charge: in fact, there are about 6,250,000,000,000,000,000 electrons in one coulomb. But don't worry, it's not a number you have to remember — you don't even have to think about electrons and coulombs because the concept of electricity, as far as we're concerned, is not about electron flow, or volumes of electrons, but about flow rate and flow pressure. And as you'll now see, electricity flow rate and pressure are given their own names which — thankfully — don't even refer to electrons or coulombs.

Going back to the water and pipe analogy, flow rate would be measured as a volume of water which flowed through the pipe during a defined period of time, say 10 litres in one minute, 1,000 litres in one hour or one litre in one second.

With electricity, flow rate is measured in a similar way, as a volume which flows past a point, during a defined period of time, except that volume is, of course, in coulombs. So, we could say that a flow rate of electricity is 10 coulombs in one minute, 1,000 coulombs in one hour or one coulomb in one second.

We could say that, but we don't! Instead, in electricity, flow rate is called current (and given the symbol I, when drawn in a diagram).

Electric current is measured in amperes (shortened to amps, or even further shortened to the unit: A), where one amp is defined as a quantity of one coulomb passing a point in one second.

Instead of saying 10 coulombs in one minute we would therefore say:

$$\frac{10}{60} \text{ coulombs per second} = 0.166 \text{ A}$$

Similarly, instead of a flow rate of 1,000 coulombs in one hour, we say:

$$\frac{1000}{3600} \text{ coulombs per second} = 0.3 \text{ A}$$

# Starting electronics

The other important thing we need to know about electricity is flow pressure. Back to our analogy with water and pipe, Figure 1.2 shows a header tank of water at a height, h, above the pipe. Water pressure is often classed as a head of water, where the height, h, in metres, is the head. The effect of gravity pushes down the water in the header tank, forming a flow pressure, forcing the water out of the pipe. It's the energy contained in the water in the header tank due to its higher position — its potential energy — which defines the water pressure.

With electricity the flow pressure is defined by the difference in numbers of electrons between two points. We say that this is a potential difference, partly because the difference depends on the positions of the points and how many electrons poten-

**Figure 1.2    A header tank's potential energy forces the water with a higher pressure**

tially exist. Another reason for the name potential difference comes from the early days in the pioneering of electricity, when the scientists of the day were making the first batteries. Figure 1.3 shows the basic operating principle of a battery, which simply generates electrons at one terminal and takes in electrons at the other terminal. Figure 1.3 also shows how the electrons from the battery flow around the circuit, lighting the bulb on their way round.

Under the conditions of Figure 1.4 (over), on the other hand, nothing actually happens. This is because the two terminals aren't joined and so electrons can't flow. (If you think about it, they are joined by air, but air is an example of a material which doesn't allow electrons to flow through it under normal conditions. Air is an insulator or a non-conductor.) Nevertheless the battery has the potential to light the bulb and so

BATTERY GENERATES ELECTRONS AT ONE
TERMINAL, AND TAKES IN ELECTRONS AT
THE OTHER TERMINAL

BULB LIGHTS UP AS
ELECTRONS FLOW
AROUND THE
CIRCUIT

ELECTRONS FLOW AROUND THE
CIRCUIT DUE TO POTENTIAL
DIFFERENCE OF BATTERY

Figure 1.3    A battery forces electrons around a circuit, only when the circuit is complete. If it's not connected (see Figure 1.4 over), no electrons flow – but it still has the potential to make them flow

# Starting electronics

EVEN THOUGH BATTERY IS NOT
CONNECTED TO THE BULB, IT
STILL HAS THE POTENTIAL TO
LIGHT THE BULB

BULB IS NOT LIT UP AS
NO ELECTRONS FLOW

THIS LEAD IS NOT CONNECTED
SO NO ELECTRONS CAN FLOW

**Figure 1.4** Even when the battery is disconnected and electrons do not flow, the battery still has a potential difference

the difference in numbers of electrons between two points (terminals in the case of a battery) is known as the potential difference. A more usual name for potential difference, though, is voltage, shortened to volts, or even the symbol V. Individual cells are rated in volts and so a cell having a voltage of 3 V has a greater potential difference than a cell having a voltage of 2 V. The higher the voltage, the harder a cell can force electrons around a circuit. Voltage is simply a way of expressing electrical pushing power.

## Relationships

You'd be right in thinking that there must be some form of relationship between this pushing power in volts and the rate of electron flow in amps. After all, the higher the voltage, the more pushing power the electrons have behind them so the faster they should flow. The relationship was first discovered

12

by a scientist called Ohm, and so is commonly known as Ohm's law. It may be summarised by the expression:

$$\frac{V}{I} = \text{a constant}$$

where the constant depends on the substance through which the current flows and the voltage is applied across. Figure 1.5 gives an example of a substance which is connected to a cell. The cell has a voltage of 2 V, so the voltage applied across the substance is also 2 V. The current through the substance is, in this case, 0.4 A. This means, from Ohm's law, that the constant for the substance is:

$$\frac{2}{0.4} = 5$$

The constant is commonly called the substance's resistance (because it is, in fact a measure of the amount the substance resists the flow of current through it) and is given the unit: $\Omega$ (pronounced ohm — not omega — after the scientist, not the Greek letter its symbol is borrowed from). So, in our example of Figure 1.5, the resistance of the substance is 5 $\Omega$. In some literature the letter R is used instead of $\Omega$. Different substances may have different resistances and may therefore change the current flowing.

CURRENT FLOWS AROUND CIRCUIT AND THROUGH SUBSTANCE (0.4 A)

CELL VOLTAGE 2V

SUBSTANCE, WHICH HAS RESISTANCE GIVEN BY OHM'S LAW:

$$= \frac{2}{0.4} = 5\,\Omega$$

**Figure 1.5** Cell's voltage is 2 V, and a current of 0.4 A flows

13

# Starting electronics

Indeed, this is so important, let's recap it and see what it all
means:

If a voltage (V — measured in volts) is applied across a resistance (R — measured in ohms), a current (I — measured
in amps) will flow. The voltage, current and resistance are
related by the expression (1):

$$\frac{V}{I} = R \tag{1}$$

The importance of this is that the current which flows depends
entirely on the values of the resistance and the voltage. The
value of the current may be determined simply by rearranging
expression 1, so that it gives (2):

$$\frac{V}{R} = I \tag{2}$$

So, a voltage of say 10 V, applied across a resistance of 20 Ω,
produces a current of:

$$\frac{10}{20} = 0.5 \text{ A}$$

Similarly, if we have a resistance, and a current is made to

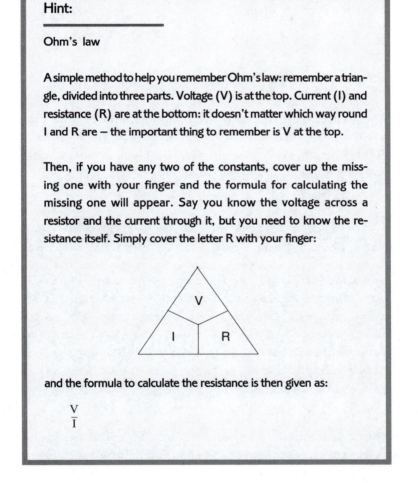

**Hint:**

Ohm's law

A simple method to help you remember Ohm's law: remember a triangle, divided into three parts. Voltage (V) is at the top. Current (I) and resistance (R) are at the bottom: it doesn't matter which way round I and R are — the important thing to remember is V at the top.

Then, if you have any two of the constants, cover up the missing one with your finger and the formula for calculating the missing one will appear. Say you know the voltage across a resistor and the current through it, but you need to know the resistance itself. Simply cover the letter R with your finger:

and the formula to calculate the resistance is then given as:

$$\frac{V}{I}$$

flow through it, then a voltage is produced across it. The value of the voltage may be determined by again rearranging expression 1, so that it now gives (3):

$$V = IR \qquad\qquad (3)$$

Thus, a current of, say, 1 A flowing through a resistance of 5 Ω, produces a voltage of:

$$1 \times 5 = 5 \text{ V}$$

across the resistance.

These three expressions which combine to make Ohm's law are the most common ones you'll ever meet in electronics, so look at 'em, read 'em, use 'em, learn 'em, inwardly digest 'em — just don't forget 'em. Right? Right.

---

**Take note — Take note — Take note — Take note**

And another thing. See the way we've said throughout, that a voltage is applied or produced across a resistance. Similarly a current flows through a resistance. Well let's keep it like that! Huh? Just remember that a voltage is across: a voltage does not flow through. Likewise, a current flows through: it is not across.

There is no such thing as a flow of voltage through a resistance, and there's no such thing as a current across a resistance.

---

## Electronic components

The fact that different resistances produce different currents if a voltage is applied across them, or produce different voltages if a current is applied through them, is one of the most useful facts in electronics.

In electronics, an amp of current is very large — usually we only use much smaller currents, say, a thousandth or so of an amp. Sometimes we even use currents smaller than this, say, a millionth of an amp! Similarly, we sometimes need only small voltages, too.

Resistances are extremely useful in these cases, because they can be used to reduce the current flow or the voltage produced across them, due to the effects of Ohm's law. We'll look at ways and means of doing this in the next chapter. All we need to know for now is that resistances are used in electronics to control current and voltage.

Table 1.1 shows how amps are related to the smaller values of current. A thousandth of an amp is known as a milliamp (unit: mA). A millionth of an amp is a microamp (unit: $\mu$A).

| Current name | Meaning | Value | Symbol |
| --- | --- | --- | --- |
| amp | — | $10^0$ A | A |
| milliamp | one thousandth of an amp | $10^{-3}$ A | mA |
| microamp | one millionth of an amp | $10^{-6}$ A | $\mu$A |
| nanoamp | one thousand millionth of an amp | $10^{-9}$ A | nA |
| picoamp | one million millionth of an amp | $10^{-12}$ A | pA |
| femtoamp | one thousand million millionth of an amp | $10^{-15}$ A | fA |

Table 1.1   Comparing amps with smaller values of current

Even smaller values of current are possible: a thousand mil-lionth of an amp is a nanoamp (unit: nA); a million millionth is a picoamp (unit: pA). Chances are, you will never use or even specify a current value smaller than these, and you will rarely even use picoamp. Milliamps and microamps are quite commonly used, though.

It's easy to move from one current value range to another, simply by moving the decimal point one way or the other by the correct multiple of three decimal places. In this way, a current of 0.01 mA is the same as a current of $10\,\mu$A which is the same as a current of 10,000 nA and so on.

Table 1.2 shows, similarly, how volts are related to smaller values of voltage. Sometimes, however, large voltages exist (not so much in electronics, but in power electricity) and so these have been included in the table. The smaller values cor-respond to those of current, that is, a thousandth of a volt is a millivolt (unit: mV), a millionth of a volt is a microvolt (unit: $\mu$V) and so on — although anything smaller than a millivolt is, again, only rarely used.

| Voltage name | Meaning | Value | Symbol |
|---|---|---|---|
| megavolt | one million volts | $10^6$V | MV |
| kilovolt | one thousand volts | $10^3$V | kV |
| volt | — | $10^0$V | V |
| millivolt | one thousandth of a volt | $10^{-3}$V | mV |
| microvolt | one millionth of a volt | $10^{-6}$V | $\mu$V |
| nanovolt | one thousand millionth of a volt | $10^{-9}$V | nV |

Table 1.2   Comparing volts with smaller and larger voltages

Larger values of voltage are the kilovolt, that is, one thousand volts (unit: kV) and the megavolt, that is, one million volts (unit: MV). In electronics, however, these are never used.

## Resistors

The components which are used as resistances are called, naturally enough, resistors. So that we can control current and voltage in specified ways, resistors are available in a number of values. Obviously, it would be impractical to have resistors of every possible value (for example, $1\,\Omega$, $2\,\Omega$, $3\,\Omega$, $4\,\Omega$) because literally hundreds of thousands — if not millions — of values would have to exist.

Instead, agreed ranges of values exist: and manufacturers make their resistors to have those values, within a certain tolerance. Table 1.3 (over) shows a typical range of resistor values, for example. This range is the most common. You can see from it that large values of resistors are available, measured in kilohms, that is, thousands of ohms (unit: $k\Omega$) and even megohms, that is, millions of ohms (unit: $M\Omega$). Sometimes the unit $\Omega$ (or the letter R if used) is omitted, leaving the units as just k or M.

Resistor tolerance is specified as a plus or minus percentage. A $10\,\Omega$ $\pm10\%$ resistor, say, may have an actual resistance within the range $10\,\Omega$ $-10\%$ to $10\,\Omega$ $+10\%$, that is, between $9\,\Omega$ and $11\,\Omega$.

As well as being rated in value and tolerance, resistors are also rated by the amount of power they can safely dissipate

| 1 Ω | 10 Ω | 100 Ω | 1 k | 10 k | 100 k | 1 M | 10 M |
|---|---|---|---|---|---|---|---|
| 1.2 Ω | 12 Ω | 120 Ω | 1k2 | 12 k | 120 k | 1M2 | — |
| 1.5 Ω | 15 Ω | 150 Ω | 1k5 | 15 k | 150 k | 1M5 | — |
| 1.8 Ω | 18 Ω | 180 Ω | 1k8 | 18 k | 180 k | 1M8 | — |
| 2.2 Ω | 22 Ω | 220 Ω | 2k2 | 22 k | 220 k | 2M2 | — |
| 2.7 Ω | 27 Ω | 270 Ω | 2k7 | 27 k | 270 k | 2M7 | — |
| 3.3 Ω | 33 Ω | 330 Ω | 3k3 | 33 k | 330 k | 3M3 | — |
| 3.9 Ω | 39 Ω | 390 Ω | 3k9 | 39 k | 390 k | 3M9 | — |
| 4.7 Ω | 47 Ω | 470 Ω | 4k7 | 47 k | 470 k | 4M7 | — |
| 5.6 Ω | 56 Ω | 560 Ω | 5k6 | 56 k | 560 k | 5M6 | — |
| 6.8 Ω | 68 Ω | 680 Ω | 6k8 | 68 k | 680 k | 6M8 | — |
| 8.2 Ω | 82 Ω | 820 Ω | 8k2 | 82 k | 820 k | 8M2 | — |

**Table 1.3   Typical resistor value range**

as heat, without being damaged. As you'll remember from our discussion on soldering irons earlier, power rating is expressed in watts (unit: W), and this is true of resistor power ratings, too.

As the currents and voltages we use in electronics are normally pretty small, the resistors we use also have small power ratings. Typical everyday resistors have ratings of $1/_4$W, $1/_3$W, $1/_2$W, 1 W and so on. At the other end of the scale, for use in power electrical work, resistors are available with power ratings up to and over 100 W or so.

Choice of resistor power rating you need depends on the resistor's use, but a reasonable value for electronics use is $1/_4$W. In fact, $1/_4$W is such a common power rating for a resistor that you can assume it for all the circuits in this book. If I give you a circuit to build which uses resistors of different power ratings, I'll tell you.

## Time out

That's all we're going to say about resistors here — in this chapter at least. In the next chapter, though, we'll be taking a look at some simple circuits you can build with resistors. We'll also explain how to measure electricity, with the aid of a meter, another useful tool which is so often used in electronics. But that's enough for now, you've learned a lot in only a little time.

If, on the other hand, you feel you want to test your brain a bit more, try the quiz over the page.

# Starting electronics

## Quiz

Answers at end of book

1. 100 coulombs of electricity flow past a point in an electrical circuit, in 20 seconds. The current flowing is?
   a  10 A
   b  2 A
   c  5 V
   d  5 A
   e  none of these.

2. A resistor of value 1 kΩ is placed in a simple circuit with a battery of 15 V potential difference. What is the value of current which flows?
   a  15 mA
   b  150 mA
   c  1.5 mA
   d  66.7 mA
   e  none of these

3. A voltage of 20 V is applied across a resistor of 100 Ω. What happens?
   a  a current of 0.2 A is generated across a resistor.
   b  a current of 5 A is generated across the resistor.
   c  a current of 5 A flows through the resistor.
   d  one coulomb of electricity flows through the resistor.
   e  none of these.

4. A current of 1 A flows through a resistor of 10 Ω. What voltage is produced through the resistor?
   a  10 V
   b  1 V
   c  100 V
   d  10 C
   e  none of these.

5. A nanoamp is?
   a  $1 \times 10^{-6}$A
   b  $1 \times 10^{-8}$A
   c  $1,000 \times 10^{-12}$A
   d  $1,000 \times 10^{-6}$A
   e  none of these.

6. A voltage of 10 MV is applied across a resistor of 1 MΩ. What is the current which flows?
   a  10 μA
   b  10 mA
   c  10 A
   d  10 MA
   e  none of these

# 2 On the boards

In this chapter we give you details of some easy-to-do experiments, designed to give you valuable practical experience. To perform these experiments you'll need some simple components and a couple of new tools.

The components you need are:

●     2 x 10 k resistors

●     2 x 1k5 resistors

●     2 x 150 Ω resistors

Power ratings and tolerances of these resistors are not important; just get the cheapest you can find.

The tools, on the other hand:

- a breadboard (such as the one we use)

- a multi-meter (such as the multi-meter we use)

are important. They are, unfortunately, quite expensive but, looked after will last you a long, long time. They're worth the expense, because you'll be able to use them for all your experiments and projects that you do and build.

Last chapter we looked at some of the essential tools you'll need if you intend to progress very far in electronics. Bread-boards and multi-meters are two more, which are also very much essential if you're at all serious in your intent to learn about electronics.

Fortunately, all the tools we show you in this book will last for years if properly treated, so even though it may seem like a lot of weeks' pension money now, it's money well spent as it's well worth getting the best you can afford!

## All aboard

A breadboard is extremely useful. With a breadboard you can construct circuits in a temporary form, changing components if required, before committing them to a permanent circuit board. This is of most benefit if you are designing the circuit from scratch and have to change components often.

If you are following a book like *Starting Electronics*, however, a breadboard is even more useful. This is because the many circuits given in the book can be built up experimentally, tested, then dismantled, so that the components may be used again and again. I'll be giving you many such experimental circuits and, although I'll also give you good descriptions of the circuits, there's nothing like building-it-yourself to find out how a circuit works. So, I recommend you get the best kind of breadboard you can find — it's worth it in the long run.

There are many varieties of breadboard. All of the better ones consist basically of a moulded plastic body which has a number of holes in the top surface, through which component leads may be easily pushed. Underneath each hole is a clip mechanism, which holds the component lead tight enough so that it can't fall out. Figure 2.1 gives the idea. The clip forms a good electrical contact, yet allows the lead to be pulled out without damage.

Generally, the clips are interconnected in groups, so that by pushing leads of two different components into two holes

Figure 2.1    The interior of a breadboard, showing the contacts

of one group you have made an electrical contact between the two leads. In this way the component leads don't have to physically touch above the surface of the breadboard to make electrical contact.

Differences lie between breadboards in the spacings and positionings of the holes, and the number of holes in each group. The majority of breadboards have hole spacings of about 2.5 mm (actually 0.1 in — which is the exact hole spacing required by a particular type of electronic component: the dual-in-line integrated circuit — I'll talk about this soon) which is fine for general-purpose use, so the only things you have to choose between are the numbers of holes in groups, the size of the breadboard and the layout (that is, where the groups are) on the breadboard.

Because there are so many different types of breadboard available, we don't specify a standard type to use in this book. So the choice of what to buy is up to you. We do, however, show circuits on a basic breadboard which is a fairly common layout. So, any circuits we show you to build on this breadboard can also be built on any similar quality breadboard, but you may have to adjust the actual practical circuit layout to suit.

Photo 2.1 shows a photograph of the breadboard we use throughout this book, in which you can see the top surface with all the component holes. Photo 2.2 shows the inside of the breadboard, with component lead clips interconnected into groups. The groups of clips are organised as two rows, the closest holes being 7.5 mm — (not just by coincidence the distance between

**Hint:**

---

The theoretical breadboard used for artwork layout purposes in this book is specified with a total of 550 contacts arranged in a main matrix of two blocks of 47 rows of five interconnected sockets, and a row of 40 interconnected sockets down each side of the main matrix. Such a theoretical block will hold upto six 14-pin or nine 8-pin dual-in-line integrated circuit packages, together with their ancillary components.

Remember though, that you may need to adapt the theoretical circuit layout to suit whichever particular breadboard you buy.

the rows of pins of a dual-in-line integrated circuit package) apart.

## ICs

Hey, wait a minute — I've mentioned a few times already this mysterious component called a dual-in-line integrated circuit, but what is it? Well, Photo 2.3 shows one in close-up while Photo 2.4 shows it, in situ, in a professional plugblock. The pins (which provide connections to the circuits integrated inside the body — integrated circuit — geddit?) are in two

# Starting electronics

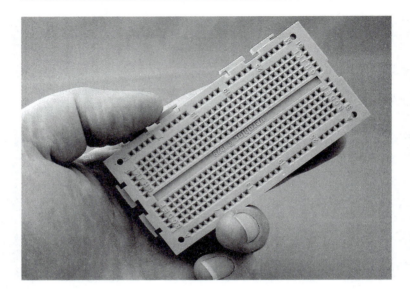

**Photo 2.1 A typical breadboard**

rows 7.5 mm apart (actually, they're exactly 0.3 in apart), so it pushes neatly into the breadboard. Because there are two rows and they are parallel — that is, in line — we call it dual-in-line (often shortened to DIL. And while we're on the subject of abbreviations, the term dual-in-line package is often shortened to DIP, and integrated circuit also is often shortened to IC).

Many types of IC exist. Most — at least as far as the hobbyist is concerned — are in this DIL form, but other shapes do exist. Often DIL ICs have different numbers of pins, e.g. 8, 14, 16, 18, 28, but the pins are always in two rows. Some of the DIL ICs with large numbers of pins have rows spaced 15 mm apart (actually, exactly 0.6 in), though.

**Photo 2.2**    Inside breadboard, showing component clips interconnected in groups

**Photo 2.3**    A DIL (dual-in-line) IC package

# Starting electronics

**Photo 2.4** **An IC mounted on a breadboard. The breadboard is designed so that an IC can be mounted without shorting the pins**

The circuits integrated inside the body of the ICs are not always the same, and so one IC can't automatically do the job of another. They need to be exactly the same type to be able to do that. This is why I always give a type number if I use an IC in an experiment. Make sure you buy the right one if you want to build an experiment, or for that matter if you ever build a project such as those you see in electronics magazines.

Once the IC is in the breadboard — in fact, once any component is in the breadboard — it's a simple matter to make connections to it by pushing in wires or other component leads to the holes and clips of the same groups.

Down the edges of the breadboard are other groups of holes connected underneath, too. These are useful to carry power supply voltages from, say, a battery, which may need to be connected into circuit at a number of points.

We can show all the various groups of holes in the bread-board block by means of the diagram in Figure 2.2, where the connected holes are shown joined by lines. This type of diagram, incidentally, will be used throughout this book to

**Figure 2.2   A breadboard pattern showing graphically the internal contacts**

show how the experimental circuits we look at are built using breadboard block. Obviously, any circuit may be built in a lot of different ways and so you don't have to follow my diagrams, or use the same breadboard as used here, but doing so will mean that your circuit is the same as mine and so easier to compare. The choice is yours. And — remember — the big advantage about using breadboard is that the components can be pulled out when the circuit is finished and you can use them again (provided you've been careful and haven't damaged them).

## The first circuit

We've done a lot of talking up to now, and not much doing, but now it's time to use your breadboard to build your first circuit. Well, to be truthful it's not really a circuit — it's just a single resistor stuck into the breadboard so that we can experiment with it.

The experiments in this chapter are all pretty simple ones, measuring the resistances of various resistors and their associated circuits. But to measure the resistances we need the other essential tool I mentioned earlier — the multi-meter. Strictly speaking a multi-meter isn't just a tool used in electronics, it's a complete piece of equipment. It can be used not only to measure resistance of resistors, but also voltage and current in a circuit. Indeed, some expensive multi-meters may be used to measure other things, too. However, you don't need an expensive one to measure only the essentials (and some non-essentials, too).

## Hint:

The multi-meter you buy and use is not important – as long as it meets a certain specification it will do the job nicely. This specification is below.

Note that any modern multi-meter should meet this specification. The specification represents just the absolute minimum you should check for, and was originally drawn up for use when buying an analogue multi-meter (ie, one with a pointer). Most modern multi-meters are of a digital nature (ie, with a digital readout) and so will usually greatly exceed the minimum specification.

While it's impossible for me to comment on how you intend using your multi-meter, so it's impossible for me to tell you which one to buy. On the other hand, it is possible for me to recommend a few specifications which you should try to match or better, when you buy your multi-meter. This is simply to ensure that your multi-meter will be as general-purpose as possible, and will perform measurements for you long after you progress from being a beginner in electronics to being an expert. The important points to remember are:

● it must have a sensitivity of at least $20\,\mathrm{k\Omega\,V^{-1}}$ on d.c. ranges. (d.c. stands for direct current).

● it must have an accuracy of no worse than ±5%.

● its smallest d.c. voltage range should be no greater than 1 V.

**Photo 2.5    A multi-meter – the one used throughout this book – although any multi-meter with at least the specification given will do**

●     its smallest current range should be no greater than $500\,\mu A$.

●     it should measure resistance in at least three ranges.

In practice, just about any modern multi-meter will meet and exceed this specification. Only older style analogue multi-meters may fall below it — digital multi-meters almost always exceed it.

Using, a multi-meter is fairly simple. It will probably have a switch on the front, which turns so that you may select which range of measurement you want. When you have connected the multi-meter up to the circuit you wish to measure (a pair of leads should be supplied with the multi-meter) the read-out will display the measurement or (on an older analogue multi-meter), the pointer of the multi-meter moves and you can read-off the measured value on the scale underneath the pointer. At the ends of the multi-meter leads are probes which allow you to connect the multi-meter to the circuit in question.

## Experiment

Using the multi-meter in our first experiment — to measure a resistor's resistance — we will now go through the procedure step-by-step, so that you get the hang of it.

The circuit built on breadboard is shown in Figure 2.3. Being only one resistor it's an extremely simple circuit. So simple that we are sure you would be able to do-it-yourself without our aid, but we might as well start off on a good footing and do the job properly — some of the circuits we'll be looking at in following chapters will not be so simple.

If you have an analogue multi-meter, you have to adjust it so that the reading is accurate. Step-by-step, this is as follows:

1)      Turn the switch to point to a resistance range (usually marker OHM ×1K, or similar).

2)      Touch the multi-meter probes together — the pointer should swing around to the right.

3)      Read the resistance scale of the multi-meter — the top one on our multi-meter marked OHMS, where the pointer crosses it — it should cross exactly on the number 0.

TEST PROBES

METER SET TO
Ω x1K RANGE

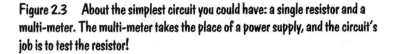

**Figure 2.3    About the simplest circuit you could have: a single resistor and a multi-meter. The multi-meter takes the place of a power supply, and the circuit's job is to test the resistor!**

4)    If it doesn't cross at 0, adjust the multi-meter using the zero adjust knob (usuallu marked 0ΩADJ).

What you've just done is the process of zeroing the multi-meter. You have to zero the multi-meter every time you use it to measure resistance. You also have to do it if you change resistance ranges. On the other hand, you never have to do it if you use your multi-meter to measure current or voltage, only resistance, or if you have a digital multi-meter.

You see, measurement of resistance relies on the voltages of cells or a battery inside the multi-meter. If a new cell is in operation, the voltage it produces may be, say, 1.6 V. But as it gets older and starts to run down, the voltage may fall to, say, 1.4 V or even lower. The zero adjustment allows you to take this change in cell or battery voltage into account and therefore make sure your resistance measurement is correct. Clever, eh?

Measurement of ordinary current and voltage, on the other hand, doesn't rely on an internal cell or battery at all, so zero adjustment is not necessary.

Now let's get back to our experiment. Following the diagram of Figure 2.3:

1)    Put a 10 k resistor (brown, black, orange bands) into the breadboard.

2)    Touch the multi-meter leads against the leads of the resistor (it doesn't matter which way round the multi-meter leads are).

3)    Read-off the scale at the point where the pointer crosses it. What does it read? It should be 10.

But how can that be? It's a 10 k resistor, isn't it? Well, the answer's simple. If you remember, you turned the multi-meter's range switch to OHM ×1K, didn't you? Officially, this should be OHM ×1k, that is, a lower case k. This tells you that whatever reading you get on the resistance scale you multiply by 1 k, that is 1000. So the multi-meter reading is actually 10,000. And what is the value of the resistor in the breadboard — 10 k (or 10,000 Ω), right!

## Hint:

### Resistor colour-code

Resistance values are indicated on the bodies of resistors in one of two ways: in actual figures, or more usually by a colour code. Resistors using figures are usually high precision or high wattage types that have sufficient space on their bodies to print characters on. Colour coding, on the other hand, is the method used on the vast majority of resistors — for two reasons. First, it is easier to read when components are in place on a printed circuit board. Second, some resistors are so small it would be impossible to print numbers on them — let alone read them afterwards.

Depending on the type of resistor, the colour code can be made up of four or five bands printed around the resistor's body (as shown below). The five-band code is typically used on more accurate resistors as it provides a more precise representation of value. Usually, the four-band code is adequate for most general purposes and it's the one you'll nearly always use — but you still need to be aware of both! Table 2.1 shows both resistors and lists the colours and values associated with each band of both four-band and five-band colour codes.

**Hint:**

The bands grouped together indicate the resistor's resistance value, while the single band indicates its tolerance.

The first band of the group indicates the resistor's first figure of its value. The second band is the second figure. Then, for a four-band coded resistor the third band is the multiplier. For a five-band coded resistor the third band is simply the resistor's third figure, while the fourth band is the multiplier. For both, the multiplier is simply the factor by which the first figure should be multiplied by (or simply the number of noughts to add) to obtain the actual resistance.

As an example, take a resistor coded red, violet, orange, silver. Looking at Table 2.1, we can see that it's obviously a four-band colour coded resistor, and its first figure is 2, second is 7, multiplier is x1000, and tolerance is ±10%. In other words, its value is 27,000 Ω, or 27 k.

# Starting electronics

**FOUR BAND RESISTORS**

1 2 3 4

| Colour | Band 1<br>1st Figure | Band 2<br>2nd Figure | Band 3<br>Multiplier | Band 4<br>Tolerance |
|---|---|---|---|---|
| Black | 0 | 0 | x1 | - |
| Brown | 1 | 1 | x10 | 1% |
| Red | 2 | 2 | x100 | 2% |
| Orange | 3 | 3 | x1000 | - |
| Yellow | 4 | 4 | x10,000 | - |
| Green | 5 | 5 | x100,000 | - |
| Blue | 6 | 6 | x1,000,000 | - |
| Violet | 7 | 7 | - | - |
| Grey | 8 | 8 | - | - |
| White | 9 | 9 | - | - |
| Gold | - | - | x0.1 | 5% |
| Silver | - | - | x0.01 | 10% |
| None | - | - | - | 20% |

**FIVE BAND RESISTORS**

1 2 3 4 5

| Colour | Band 1<br>1st Figure | Band 2<br>2nd Figure | Band 3<br>3rd Figure | Band 4<br>Multiplier | Band 5<br>Tolerance |
|---|---|---|---|---|---|
| Black | 0 | 0 | 0 | x1 | - |
| Brown | 1 | 1 | 1 | x10 | 1% |
| Red | 2 | 2 | 2 | x100 | 2% |
| Orange | 3 | 3 | 3 | x1000 | - |
| Yellow | 4 | 4 | 4 | x10,000 | - |
| Green | 5 | 5 | 5 | x100,000 | 0.5% |
| Blue | 6 | 6 | 6 | x1,000,000 | 0.25% |
| Violet | 7 | 7 | 7 | x10,000,000 | 0.1% |
| Grey | 8 | 8 | 8 | - | 0.01% |
| White | 9 | 9 | 9 | - | - |
| Gold | - | - | - | x0.1 | 5% |
| Silver | - | - | - | x0.01 | 10% |

**Table 2.1    Resistor colour code**

40

In practice, you might find that your resistor's measurement isn't exactly 10 k. It may be, say, 9.5 k or 10.5 k. This is due, of course, to tolerance. Both the resistor and the multi-meter have a tolerance: indicated on the resistor by the last coloured band: the multi-meter's is probably around ± 5%. Chances are, though, you'll find the multi-meter reading is as close to 10 k as makes no difference.

Now you've seen how your multi-meter works, you can use it to measure any other resistors you have, if you wish. You'll find that lower value resistors need to be measured with the range switch on lower ranges, say $\Omega$ x 100. Remember — if you have an analogue multi-meter — every time you intend to make a measurement you must first zero the multi-meter. The process may seem a bit long-winded for the first two or three measurements, but after that you'll get the hang of it.

## The second circuit

Figure 2.4 shows the next circuit we're going to look at and how to build it on breadboard. It's really just another simple circuit, this time consisting of two resistors in a line — we say they're in series. The aim of this experiment is to measure the overall resistance of the series resistors and see if we can devise a formula which allows us to calculate other series resistors' overall resistances without the need of measurement.

Figure 2.5 shows the more usual way of representing a circuit in a drawing — the circuit diagram. What we have done is

# Starting electronics

**Figure 2.4    Two resistors mounted on the breadboard in series**

replace the actual resistor shapes with symbols. Resistor symbols are zig-zag lines usually, although sometimes small oblong boxes are used in circuit diagrams. The resistors in the circuit diagram are numbered R1 and R2, and their values are shown, too.

Meters in circuit diagrams are shown as a circular symbol, with an arrow to indicate the pointer. To show it's a resistance multi-meter (that is, an ohm-multi-meter, more commonly

**Figure 2.5    A circuit diagram of two resistors in series. The meter is represented by a round symbol**

42

called just ohmmeter) the letter R is shown inside it. While we're on the topic of circuit diagram symbols, Figure 2.6 shows a few very common ones (including resistor and meter) which we'll use in this book. Look out for them later!

You should have noticed that there is no indication of the breadboard in the circuit diagram of Figure 2.5. There is no need. The circuit diagram is merely a way of showing components and their electrical connections. The physical connection details are in the breadboard layout diagram of Figure 2.4. From now on, we'll be using two such diagrams with every new circuit. If you're feeling particularly adventurous you might care to build your own circuit on breadboard, following only the circuit diagram — not the associated breadboard layout. It doesn't matter if your circuit has a different layout to ours, it will still work as long as all the electrical connections are there.

Figure 2.6    Commonly used symbols

Back to the circuit: it's now time to measure the overall resistance of the series resistors. Following the same instructions we gave you before, do it!

If your measurement is correct you should have a reading of 20 k. But what does this prove? Well, it suggests that there is a relationship between the separate resistors (each of value 10 k) and the overall resistance. It looks very much as though the overall resistance (which we call, say, $R_{OV}$) equals R1 + R2. Or put mathematically:

$$R_{OV} = R1 + R2$$

But how can we test this? The easiest way is to change the resistors. Try doing the experiment with two different resistors. You'll find the same is true: the overall resistance always equals the sum of the two separate resistances.

By experiment, we've just proved the law of series resistors. And it doesn't just stop at two resistors in series. Three, four, five, in fact, any number of resistors may be in series — the overall resistance is the sum of the individual ones. This can be summarised mathematically as:

$$R_{OV} = R1 + R2 + R3 + ...$$

Try it yourself!

## The next circuit

There is another way two or more resistors may be joined. Not end-to-end as series joined resistors are, but joined at both ends. We say resistors joined together at both ends are in parallel. Figure 2.7 shows the circuit diagram of two resistors joined in parallel, and Figure 2.8 shows a breadboard layout. Both these resistors are, again, 10 k resistors. What do you think the overall resistance will be? It's certainly not 20 k!

Measure it yourself using your multi-meter and breadboard.

You should find that the overall resistance is 5 k. Odd, eh? Replace the two 10 k resistors with resistors of different value say, two 150 Ω resistors (brown, green, brown). The overall resistance is 75 Ω.

**Figure 2.7    The circuit diagram for two resistors in parallel, with the meter symbol**

So, we can see that if two equal value resistors are in parallel, the overall resistance is half the value of one of them. This is a quite useful fact to remember when two parallel resistors are equal in value, but what happens when they're not?

**Figure 2.8** The two parallel resistors shown in the breadboard, with the meter in place to test their combined resistance

Try the same circuit, but with unequal resistors this time, say, one of 10k and the other of 1k5 (brown, green, red — shouldn't you be learning the resistor colour code?). What is the overall resistance? You should find it's about 1k3 — neither one thing nor the other! So, what's the relationship?

Well, a clue to the relationship between parallel resistors comes from the fact that, in a funny sort of way, parallel is the inverse of series. So if we inverted the formula for series resistors we saw earlier:

$$R_{OV} = R1 + R2 + R3 + ...$$

we would get:

$$\frac{1}{R_{ov}} = \frac{1}{R1} + \frac{1}{R2} + \frac{1}{R3} + \ldots$$

and this is the formula for parallel resistors. Let's try it out on the resistors of this last experiment. Putting in the values, 10 k and 1k5 we get:

$$\frac{1}{R_{ov}} = \frac{1}{10,000} + \frac{1}{1500}$$

$$\text{which is:} = \frac{1500 + 10,000}{15,000,000}$$

$$= \frac{11,500}{15,000,000}$$

$$= 0.00076$$

$$\text{So:} \quad R_{ov} = 1304 \ \Omega$$

which is about 1k3, the measured value.

This is the law of parallel resistors, every bit as important as that of series resistors. Remember it!

If there are only two resistors in parallel, you don't have to calculate it the way we've just done here — there is a simpler way, given by the expression:

$$R_{ov} = \frac{R1 \times R2}{R1 + R2}$$

But if there are three or more resistors in parallel you have to use the long method, I'm afraid.

# Starting electronics

## More and more complex circuits

Circuits we've looked at so far have been very simple, really. Much more complex ones await us over the coming chapters. We'll also be introduced to several new components, such as capacitors, diodes, transistors, and ICs. Fortunately, most of the new circuits and components follow these basic laws we've studied, so you'll be able to understand their operation without much ado.

The aim of these basic laws is to simplify complex circuits so that we may understand them and how they work, with reference to the circuits we have already seen. For example, the circuit in Figure 2.9 is quite complex. It consists of many resistors in a network. But by grouping the resistors into smaller circuits of only series and parallel resistors, it's possible to simplify the whole circuit into one overall resistance.

In fact, I'm going to leave you with that problem now. What you have to do, is to calculate the current, I, which flows from the battery. You can only do that by first finding the circuit's

48

**Figure 2.9    A comparatively complex circuit showing a number of resistors in parallel and in series. By breaking the circuit down into groups, the overall resistance can be calculated**

overall resistance. If you can't do it by calculation you can always do it by building the circuit up on breadboard and measuring it. Resistor and multi-meter tolerances, of course, will mean the measured result may not give exactly the same result as the calculated one.

Meanwhile, there's always the quiz over the page to keep you occupied!

# Starting electronics

## Quiz

Answers at end of book

1 Five 10 kΩ resistors are in series. One 50 kΩ resistor is placed in parallel across them all. The overall resistance is:
a 100 kΩ
b 52 kΩ
c 25 kΩ
d 50 kΩ
e None of these.

2 Three 30 kΩ resistors are in parallel. The overall resistance is:
a 10 kΩ
b 90 kΩ
c 27 kΩ
d 33 kΩΩ
e 7k5
f None of these.

3 When must you zero a multimeter?
a Whenever a new measurement is to be taken.
b Whenever you turn the range switch.
c Whenever you alter the zero adjust knob.
d Whenever a resistance measurement is to be made.
e All of these.
f None of these.

4 The law of series resistors says that the overall resistance of a number of resistors in series is the sum of the individual resistances.

True or false?

5 The overall resistance of two parallel resistors is 1 kΩ. The individual resistance of these resistors could be:
a 2 kΩ and 2 kΩ
b 3 kΩ and 1k5
c 6 kΩ and 1k2
d 9 kΩ and 1125 Ω
e a and b
f All of these.
g None of these.

50

# 3   Measuring current and voltage

There's not too much to buy for this chapter of *Starting Electronics*. As usual there's a small number of resistors:

● 2 x lk5

● 1 x 4k7

● 2 x 100 k

As before, power ratings and tolerance of these resistors are not important.

We're looking at current and voltage in our experiments in this chapter, so you'll need to have a voltage and current source. The easiest and cheapest method is a simple battery. PP3, PP6, or PP9 sizes are best — as we need a 9 V source.

Whichever battery you buy, you'll also need its corresponding battery connectors. These normally come as a pair of push-on connectors and coloured connecting leads. The red lead connects to the positive battery terminal; the black lead connects to the negative terminal.

The ends of the connecting leads furthest from the battery are normally stripped of insulation for about the last few millimetres or so, and tinned. Tinning is the process whereby the loose ends of the lead are soldered together. If the leads are tinned you'll be able to push them straight into your breadboard. If the leads aren't tinned, on the other hand, don't just push them in because the individual strands of wire may break off and jam up the breadboard. Instead, tin them yourself using your soldering iron and some multicored solder. The following steps should be adhered to:

1)      twist the strands of the leads between your thumb and first finger, so that they are tightly wound, with no loose strands,

2)      switch on the soldering iron. When it has heated up, tin the iron's tip with multicore solder until the end is bright and shiny with molten solder. If necessary, wipe off excess solder or dirt from the tip, on a piece of damp sponge; keep a small piece of sponge just for this purpose,

3)      apply the tip of the iron to the lead end, and when the wires are hot enough apply the multicore solder. The solder should flow over the wires smoothly. Quickly remove the tip and the multicore solder, allowing the lead to cool naturally (don't blow on it as the solder may crack if it cools too quickly). The lead end should be covered in solder, but no excess solder should be present. If a small blob of solder has

formed on the very end of the lead, preventing the lead from being pushed into your breadboard, just cut it off using your side cutters.

We'll start our study, this chapter, with a brief recap of the ideas we covered in the last chapter. We saw then that resistors in series may be considered as a single equivalent resistor, whose resistance is found by adding together the resistance of each resistor. This is called the law of series resistors, which is given mathematically by:

$$R_{OV} = R1 + R2 + R3 + \dots$$

Similarly, there is a law of parallel resistors, by which the single equivalent resistance of a number of resistors connected in parallel is given by:

$$\frac{1}{R_{OV}} = \frac{1}{R1} + \frac{1}{R2} + \frac{1}{R3} + \dots$$

Using these two laws many involved circuits may be broken down, step by step, into an equivalent circuit consisting of only one equivalent resistor. The circuit in Figure 2.9 last chapter was one such example and, if you remember, your homework was to calculate the current I from the battery. To do this you first have to find the single equivalent resistance of the whole network, then use Ohm's law to calculate the current.

Figure 3.1 shows the first stage in tackling the problem, by dividing the network up into a number of smaller networks. Using the two expressions above, associated with resistors in series and parallel, we can start to calculate the equivalent resistances of each small network as follows:

# Starting electronics

**Figure 3.1** A resistor network in which the total resistance can only be calculated by breaking it down into blocks

## Network A

Network A consists of two resistors in parallel. The equivalent resistance, $R_A$, may therefore be calculated from the above expression for parallel resistors. However, we saw in the last chapter that the resistance of only two parallel resistors is given by the much simpler expression:

$$R_A = \frac{R2 \times R3}{R2 + R3}$$

which gives:

$$R_A = \frac{20k \times 12k}{32k}$$

$$= 7k5$$

## Network B

Three equal, parallel resistors form this network. Using the expression for parallel resistors we can calculate the equivalent resistance to be:

$$\frac{1}{R_B} = \frac{1}{15k} + \frac{1}{15k} + \frac{1}{15k} + \dots$$

which gives:

$$R_B = 5k$$

### Hint:

This is an interesting result, as it shows that the equivalent resistance of a number of equal, parallel resistors may be easily found by dividing the resistance of one resistor, by the total number of resistors. In this case we had three, 15 k resistors, so we could simply divide 15 k by three to obtain the equivalent resistance. If a network has two equal resistors in parallel, the equivalent resistance is one half the resistance of one resistor (that is, divide by 2). If four parallel resistors form a network, the equivalent resistance is a quarter (that is, divide by 4) the resistance of one resistance of one resistor. Hmm, very useful. Must remember that — right?

## Network C

Using the simple expression for two, unequal parallel resistors:

$$R_C = \frac{R4 \times R5}{R4 + R5}$$

$$= 6k$$

## Network D

The overall resistance of two series resistors is found by adding their individual resistances. Resistance $R_D$ is therefore 10 k.

We can now redraw the whole network using their equivalent resistances, as in Figure 3.2, and further simplify the resultant networks.

**Figure 3.2** **The same network, broken down into smaller networks using equivalent resistances (networks E, F and G)**

## Network E

This equivalent resistance is found by adding resistances $R_A$ and $R_B$. It is therefore 12k5.

## Network F

Same as network E. Resistance $R_F$ = 16 k.

## Network G

Two equal parallel resistors, each of 10 k. Resistance $R_G$ is therefore 5 k. Figure 3.3 shows further simplification.

**Figure 3.3** **The network further simplified (network H) into two unequal parallel resistances**

## Network H

Two unequal, parallel resistors. Resistance $R_H$ is therefore given by:

$$R_H = \frac{R_E \times R_F}{R_E + R_F}$$

$$= \frac{12k5 \times 16k}{12k5 + 16k}$$

$$= 7k02$$

Figure 3.4 shows a further simplification.

**Figure 3.4    The network simplified into a string of series resistances (network I)**

## Network I

Resistance $R_I$ is found by adding resistances of resistors $R_1$, $R_H$ and $R_G$. Resistance $R_I$ is therefore 22k02.

Figure 3.5 shows the next stage of simplification with two unequal parallel resistors. The whole network's equivalent resistance is therefore given by:

$$R_N = \frac{R_I \times R_{12}}{R_I + R_{12}}$$

$$= 20k$$

Now, from Ohm's law we may calculate the current I, flowing from the battery, using the expression:

$$I = \frac{V}{R}$$

$$= \frac{10}{20k}$$

$$= 0.5 \times 10^{-3} A = 0.5mA$$

Did you get the same answer?

**Figure 3.5    The final simplification, into two parallel resistances. From this the current can be calculated**

## One man's meter ....

Of course, there is another way the current I may be found. Instead of calculating the equivalent resistance of the network you could measure it using your multi-meter. But first you would need to obtain resistors of all the values in the network. Then you must build the network up on breadboard, with those resistors.

Take note — Take note — Take note — Take note

The measured result may not, unfortunately, be exactly the same as the calculated result — due to resistor tolerances and multi-meter tolerance.

Finally, from this measured result and Ohm's law, the current I may be calculated.

Hint:

Your multi-meter can be used directly to measure current (it's a multi-meter, remember) so after building up your network of resistors you can simply read off the current flowing. Figure 3.6 illustrates how a multi-meter should be connected in a circuit to allow it to indicate the current through the circuit. Note the use of the words in and through. In chapter 1 we looked closely at current and saw that it is a flow of electrons around a circuit. To measure the current we must position the multi-meter in the circuit — in other words the current going around the circuit must also go through the multi-meter. This is an important point when measuring current. Remember it!

# Measuring current and voltage

**Figure 3.6** **A diagram showing a meter connected in circuit, with the red lead at the point of higher potential**

Take note — Take note — Take note — Take note

If you are using an analogue multi-meter, make sure you have your leads the correct way round. When the current through the multi-meter flows in the right direction the movement and pointer turn clockwise. In the wrong direction on the other hand, the movement and pointer attempts to turn anti-clockwise. At best you won't get a valid measurement of current if the multi-meter is the wrong way round — at worst you'll damage the movement.

A digital multi-meter, however, will merely give a negative reading.

But what is the right way round? And how do you tell the difference? The answers are quite simple really; both given by the fact that your multi-meter has two leads, one red, one black. The colour coding is used to signify which lead to connect to which point in a circuit. By convention, black is taken to be the colour signifying a lower potential. Red, also by convention, signifies a higher potential. So, the multi-meter should be connected into the circuit with its red lead touching the point of higher potential and the black lead touching the point of lower potential. This is illustrated in Figure 3.6 with symbols close to the multi-meter (+ for the red lead, – for the black). In the circuit shown, the point of higher potential is the side of the circuit to the positive terminal of the battery (also marked +).

It doesn't matter where in the circuit the multi-meter is placed, the red lead must always connect to the point of higher potential. Figure 3.7, for example, shows the multi-meter at a different position, but the measurement is the same and the multi-meter won't be damaged as long as the red lead is connected to the point of higher potential.

## Negative vibes

Figures 3.1 to 3.7 all show current flowing from the positive terminal of the battery to the negative terminal. Now we know that current is made up of a flow of electrons so we might assume that electrons also flow from positive to negative battery terminals. We might — but we'd be wrong, because electrons are actually negatively charged and therefore flow from negative to positive battery terminals. But a negative flow in one direction is exactly the same as a positive flow

**Figure 3.7** The meter connected to a different part of the circuit. The effect is the same as long as the red lead is correctly connected

in the opposite direction (think about it!) so the two things mean the same. However, it does mean that we must define which sort of current we are talking about when we use the term. Figure 3.8 defines it graphically: conventional current (usually called simply current) flows from positive to negative; electron current flows from negative to positive. Whenever we talk about current from now on, you should take it to mean conventional current.

## A meter with potential

Of course, your multi-meter can measure voltage, too. But how should it be connected into the circuit to do it? Well, the answer is that you don't connect it into the circuit — because

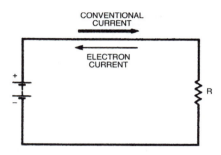

**Figure 3.8** *Conventional current travels from positive to negative; electron current travels in the opposite direction*

voltage you remember, is across a component, not through it. So, to measure the voltage across a component you must connect the multi-meter across the component, too. Figure 3.9 shows a simple example of a circuit comprising two resistors and a battery. To find the voltage across, say, the lower resistor, the multi-meter is just connected across that resistor. Like the measurement of current, the red (+) lead of the multi-meter is connected to the more positive side of the circuit and, the black (–) lead to the more negative side.

**Figure 3.9** *Meter connections to find the voltage across a single resistor*

# Measuring current and voltage

## Practically there

Now that we've looked at the theory behind current and voltage measurement, let's go on and build a few circuits to put it into practice. Figure 3.10 shows the breadboard layout of the circuit of Figure 3.6, where a single resistor is connected in series with a multi-meter and a battery. With this circuit we can actually prove Ohm's law. The procedure is as follows:

1) set the meter's range switch to a current range — the highest one, say, 10 A,

2) insert a resistor (of say 1k5) into the breadboard, and connect the battery leads,

3) touch the multi-meter leads to the points indicated.

You will probably see the pointer of the multi-meter move (but only just) as you complete step 3. The range you have set the multi-meter to is too high — the current is obviously a lot smaller than this range, so turn down the range switch (to, say, the range nearest to 250 mA) and do step 3 again.

This time the pointer should move just a bit further, but still not enough to allow an accurate reading. Turn down the range switch again, this time to, say, the 25 mA range and repeat step 3. Now you should get an adequate reading, which should be 6 mA give or take small experimental errors.

Is this right? Let's compare it to the expression of Ohm's law:

$$I = \frac{V}{R}$$

# Starting electronics

**Figure 3.10    A breadboard layout for the circuit of Figure 3.6: the meter is in series with the resistor; only the meter itself makes the circuit complete**

If we insert known values of voltage and resistance into this we obtain:

$$I = \frac{9}{1500} = 6 \times 10^{-3}A = 6mA$$

Not bad, eh? We've proved Ohm's law!

You can try this again using different values of resistor if you wish, but don't use any resistors lower than about 500 Ω as you won't get an accurate reading, because the battery can't supply currents of more than just a few mA. Even if it could the resistor would overheat due to the current flowing through it.

In all cases where the battery is able to supply the current demanded by the resistor, the expression for Ohm's law will hold true.

Take note — Take note — Take note — Take note

Make sure that you start a new measurement with the range switch set to the highest range and step down. This is a good practice to get into, because it may prevent damage to the multi-meter by excessive currents. Even though circuits we look at in *Starting Electronics* will rarely have such high currents through them, there may be a time when you want to measure an unknown current or voltage in another circuit; if you don't start at the highest range — zap — your multi-meter could be irreparably damaged.

### Hint:

The scale you must use to read any measurement from, depends on the range indicated by the meter's range switch. When the range switch points, say, to the 10 A range, the scale with the highest value of 10 should be used and any reading taken represents the current value. When the range switch points to, say, the 250 mA range, on the other hand, (and also the other two ranges 2.5 mA and 25 mA), the scale with the highest value of 25 should be used. It all sounds tricky doesn't it? Well, don't worry, once you've seen how to do it, you'll be taking measurements easily and quickly, just like a professional.

Note that current (and voltage) scales read in the opposite direction to the resistance scale we used in the last chapter, and they are linear. This makes them considerably easier to use than resistance scales and they are also more accurate as you can more easily judge a value if the pointer falls between actual marks on the scale.

## Voltages

When you measure voltages with your multi-meter the same procedure should be followed, using the highest voltage ranges first and stepping down as required. The voltages you are measuring here are all direct voltages as they are taken from a 9 V d.c. battery. So you needn't bother using the three highest d.c. voltage ranges on the multi-meter, as your 9 V battery can't generate a high enough voltage to damage the meter anyway. Also, don't bother using the a.c. voltage ranges as they're — pretty obviously — for measuring only alternating (that is, a.c.) voltages.

As an example you can build the circuit of Figure 3.9 up on your breadboard, shown in Figure 3.11. What is the measured voltage? It should be about 4.5 V.

Now measure the voltage across the other resistor — it's also about 4.5 V. Well, that figures, doesn't it? There's about 4.5 V across each resistor, so there is a total of 2 x 4.5 V that is, 9 V across them both: the voltage of the battery. This has demonstrated that resistors in series act as a voltage divider or a potential divider, dividing up the total voltage applied

RED +ve) LEAD

BLACK (-ve) LEAD

METER SET TO
MEASURE DC
VOLTAGE

**Figure 3.11    A breadboard layout for the circuit in Figure 3.9: measuring the voltage across R2. This will be the same as R1, and each will be around 4.5 V – half the battery voltage**

across them. It's understandable that the voltage across each resistor is the same and half the total voltage, because the two resistors are equal. But what happens if the two resistors aren't equal?

Build up the circuit of Figure 3.12. What is the measured voltage across resistor R2 now? You should find it's about 2.1 V. The relationship between this result, the values of the two resistors and the applied battery voltage is given by the voltage divider rule:

**Figure 3.12    A circuit with two unequal, series resistors. This is used in the text to illustrate the voltage divider rule, one of the most fundamental rules of electronics**

$$V_{out} = \frac{R2}{R1 + R2} \times V_{in}$$

where $V_{in}$ is the battery voltage and $V_{out}$ is the voltage measured across resistor R2.

We can check this by inserting the values used in the circuit of Figure 3.12:

$$V_{out} = \frac{1k5}{1k5 + 4k7} \times 9$$

$$= \frac{13500}{6200}$$

$$= 2.18V$$

In other words, close enough to our measured 2.1 V to make no difference. The voltage divider rule, like Ohm's law and the laws of series and parallel resistors, is one of the fundamental laws which we must know. So, remember it! ok?

# Measuring current and voltage

## Pot-heads

Certain types of components exist, ready-built for this voltage dividing job, known as potentiometers (commonly shortened to just pots). They consist of some form of resistance track, across which a voltage is applied, and a wiper which can be moved along the track forming a variable voltage divider. The total resistance value of the potentiometer track doesn't change, only the ratio of the two resistances formed either side of the wiper. The basic symbol of a potentiometer is shown in Figure 3.13(a).

A potentiometer may be used as a variable resistor by connecting the wiper to one of the track ends, as shown in Figure 3.13(b). Varying the position of the wiper varies the effective resistance from zero to the maximum track resistance. This is useful if we wish to, say, control the current in a particular part of the circuit; increasing the resistance decreases the current and vice versa.

**Figure 3.13    A variety of symbols used for variable resistors, or potentiometers**

These two types of potentiometer are typically used when some function of an appliance e.g., the volume control of a television, must be easily adjustable. Other types of potentiometer are available which are set at the factory upon manufacture and not generally touched afterwards e.g., a TV's height adjustment. Such potentiometers are called preset potentiometers. The only difference as far as a circuit diagram is concerned is that their symbols are slightly changed. Figures 3.13(c) and (d) show preset potentiometers in the same configurations as the potentiometers of Figures 3.13(a) and (b). Mechanically, however, they are much different.

## Meters made

The actual internal resistance of a multi-meter must be borne in mind when measuring voltages as it can affect measurements taken. We can build a circuit (Figure 3.14) which shows exactly what the effect of multi-meter resistance is. As both resistors in the circuit are equal, we can see that the measured voltage should be half the battery voltage that is, 4.5 V (use the voltage divider rule, if you don't believe it!). But when

**Figure 3.14    A circuit which is used to show the effect of the meter's own resistance in a circuit**

you apply your multi-meter across resistor R2 you find that the voltage indicated is only about 3 V.

The fact is that when the multi-meter is not connected to the circuit the voltage is 4.5 V, but as soon as the multi-meter is applied, the voltage across resistor R2 falls to 3 V. Also, the voltage across resistor R1 rises to 6 V (both voltages must add up to the battery voltage, remember). Applying the multi-meter affects the operation of the circuit, because the multi-meter resistance is in parallel with resistor R2.

## Ohms per volt

A meter's internal resistance is stated as a number of ohms per volt. For example, the resistance of the meter we use in this book is 20,000 ohms per volt (written $20\,k\Omega V^{-1}$) on d.c. voltage scales so when it's to read 4.5 V, it's resistance is 90 k. This resistance, in parallel with resistor R2 forms an equivalent resistance, given by the law of parallel resistors, of:

$$R_{eq} = \frac{100k \times 90k}{100k + 90k}$$

$$= 47k4$$

This resistance is now the new value of R2 in the circuit so, applying the voltage divider rule, the measured voltage will be:

$$V = \frac{47k4}{100k + 47k4} \times 9$$

$$= 2.9V$$

which is roughly what you measured (hopefully).

Any difference between the actual measurement and this calculated value may be accounted for because this lower voltage causes a lower multi-meter resistance which, in turn, affects the voltage which, in turn, affects the resistance, and so on until a balance is reached. All of this occurs instantly, as soon as the multi-meter is connected to the circuit.

What this tells us is that you must be careful when using your multi-meter to measure voltage. If the circuit-under-test's resistance is high, the multi-meter resistance will affect the circuit operation, causing an incorrect reading.

Any multi-meter will affect operation of any circuit to a greater or lesser extent — but the higher the multi-meter resistance compared with the circuit resistance, the more accurate the reading.

> **Hint:**
> _____
>
> A good rule-of-thumb to ensure reasonably accurate results is to make sure that the multi-meter resistance is at least ten times the circuit's resistance.

Fortunately, because of their very nature, digital multi-meters usually have an internal resistance in the order of millions of ohms; a fact which allows it to accurately measure voltages in circuits with even high resistances.

Well. We've reached the end of this chapter, so how about giving the quiz over the page a go to see if you've really learned what you've been reading about.

# Starting electronics

## Quiz

Answers at end of book

1   The voltage produced by a battery is:
    a   9 V
    b   alternating
    c   1.5 V
    d   direct
    e   none of these

2   The voltage across the two resistors of a simple voltage divider:
    a   is always 4.5 V
    b   is always equal
    c   always adds up to 9 V
    d   all of these
    e   none of these

3   The voltage divider rule gives:
    a   the current from the battery
    b   the voltage across a resistor in a voltage divider
    c   the applied voltage
    d   b and c
    e   all of these
    f   none of these

4   Conventional current flows from a point of higher electrical potential to one of lower electrical potential.

    True or false?

5   Applying a multi-meter to a circuit, to measure one of the circuit's parameters, will always affect the circuit's operation to a greater or lesser extent.

    True or false?

6   Twenty-five, 100 kΩ resistors are in parallel. What is the equivalent resistance of the network?
    a   4 kΩ
    b   2 kΩ
    c   2k0108
    d   5 kΩ
    e   none of these

# 4  Capacitors

You need a number of components to build the circuits in this chapter:

- 1 x 15 k resistor

- 1 x 22 $\mu$F electrolytic capacitor

- 1 x 220 $\mu$F electrolytic capacitor

- 1 x 470 $\mu$F electrolytic capacitor

- 1 x switch

As usual, the resistor power rating and tolerance are of no importance. The capacitor must, however, have a voltage rating of 10 V or more. The switch can be of any type, even an old light switch will do if you have one. A slide switch is ideal, however, both common and cheap.

# Starting electronics

To make the connections between the switch and your breadboard, you can use common multi-strand wire if you like, and tin the ends which push into the breadboard — see last chapter for instructions — but a better idea is to use 22 SWG single-strand tinned copper wire. This is quite rigid and pushes easily into the breadboard connectors without breaking up. It's available in reels, and a reel will last you quite a while, so it's worth buying for this and future purposes.

Take note — Take note — Take note — Take note

If short strands break off from multi-strand wire and get stuck in the breadboard connectors you might get short circuits occurring. Make sure — if you use multi-strand wire — you tin wire ends properly. Even better — don't use multi-strand wire, instead use single-strand wire.

Solder a couple of short lengths of the wire to switch connections and push on a couple of bits of insulating covering (you can buy this in reels also, but if you have a few inches of spare mains cable you'll find you can strip the few centimetres you need from this) to protect against short circuits.

## Capacitors

We're going to take a close look at a new type of component this chapter: the capacitor. We shall see how capacitors function, firstly in practical experiments which you can do for yourself and then, secondly with a bit of theory (boring, boring, boring! — Who said that?) which will show you a few things that can't easily be demonstrated in experiments.

So, let's get started. What exactly makes up a capacitor? Well, there's an easy answer: a capacitor is made up from two electrically conductive plates, separated from each other by a sandwiched layer of insulating material. Figure 4.1 gives you a rough idea.

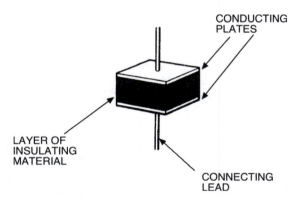

**Figure 4.1    A simple diagrammatic representation of a capacitor**

However, this is an oversimplification really; there are many, many different types of capacitor and, although their basic make-up is of two insulated plates they are not always made in quite this simple form. Nevertheless, the concept is one to be going along with.

## Down to business

Figure 4.2(a) shows the circuit symbol of an ordinary capacitor. As you'll appreciate, this illustrates the basic capacitor make-up we've just seen. Figure 4.2(b), on the other hand, shows the circuit symbol of a specific although very common type of capacitor: the electrolytic capacitor.

(a)                    (b)

**Figure 4.2   Circuit symbols for (a) an ordinary and (b) an electrolytic capacitor**

---

### Hint:

The electrolytic capacitor gets its name from the fact that its basic capacitor action is derived through the process of electrolysis when the capacitor is connected into circuit. This electrolytic action means that an electrolytic capacitor must be inserted into circuit the right way round, unlike most other capacitors. We say that electrolytic capacitors are polarised and to show this the circuit symbol has positive (white) and negative (black) plates. Sometimes as we've shown, a small positive symbol (+) is drawn beside the electrolytic capacitor's positive plate, to reinforce this.

# Capacitors

In our experiments in this chapter, we're going to use electrolytic capacitors, not because we like to be awkward, but because the values of capacitance we want are quite high. And electrolytic capacitors are generally the only ones capable of having these values, while keeping to a reasonable size and without being too expensive. Anyway, not to worry, all you have to do is remember to put the capacitors into circuit the right way round.

You can make sure of this, as all electrolytic capacitors have some kind of marking on them which identifies positive and negative plates. Figure 4.3 shows two types of fairly typical electrolytic capacitors. One, on the left, has what is called an axial body, where the connecting leads come out from each end. One end, the positive plate end, generally has a ridge around it and sometimes is marked with positive symbols (as mine is). Sometimes, the negative plate end has a black band around it, or negative symbols.

The capacitor on the right has a radial body, where both leads come out from one end. Again, however, one or sometimes both of the leads will be identified by polarity markings on the body.

Whatever type of electrolytic capacitor we actually use in circuits, we shall show the capacitor in a breadboard layout diagram as an axial type. This is purely to make it obvious (due to the ridged positive plate end) which lead is which. Both types are, in fact, interchangeable as long as the correct polarity is observed, and voltage rating (that is, the maximum voltage which can be safely applied across its leads — usually written on an electrolytic capacitor's body) isn't exceeded.

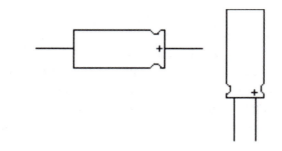

**Figure 4.3    (Left) an axial capacitor and (right) a radial capacitor**

## Measuring up

Capacitance values are measured in a unit called the farad (named after the scientist: Faraday) and given the symbol: F. We'll define exactly what the farad is later, suffice to say here that it is a very large unit. Typical capacitors have values much, much smaller. Fractions such as a millionth of a farad (that is, one microfarad: $1\,\mu F$), a thousand millionth of a farad (that is, one nanofarad: $1\,nF$), or one million millionth of a farad (that is, one picofarad: $1\,pF$) are common. Sometimes, like the $\Omega$ or R of resistances, the unit F is omitted — $15\,n$, instead of $15\,nF$, say.

The circuit for our first experiment is shown in Figure 4.4. You should see that it bears a striking resemblance to the voltage divider resistor circuits we looked at last chapter, except that a capacitor has taken the place of one of the resistors. Also included in the circuit of Figure 4.4 is a switch. In the last chapter's circuits, no switch was used as we were only interested in static conditions after the circuits were connected.

**Figure 4.4   A simple resistor and capacitor circuit**

The circuits this chapter, on the other hand, have a number of dynamic properties which occur for only a while after the circuit is connected. We can use the switch to make sure that we see all of these dynamic properties, from the moment electric current is allowed to flow into the circuit.

A breadboard layout for the circuit is shown in Figure 4.5. Make sure the switch is off before you connect your battery up to the rest of the circuit.

As the switch is off, no current should flow so the meter reading should be 0 V. If you've accidentally had the switch on and current has already flowed, you'll have a voltage reading other than zero, even after the switch is consequently turned off. In such a case, get a short length of wire and touch it to both leads of the capacitor simultaneously, for a few seconds, so that a short circuit is formed. The voltage displayed by the meter should rapidly fall to 0 V and stay there.

Now you're ready to start, but before you switch on, read the next section!

# Starting electronics

**Figure 4.5    A breadboard layout for the circuit in Figure 4.4**

## Getting results

As well as watching what happens when the circuit is switched on, you should make a record of the voltages displayed by the meter, every few seconds. To help you do this a blank table of results is given in Table 4.1. All you have to do is fill in the voltages you have measured, at the times given, into the table at successive measurement points. You'll find that the voltage increases from zero as you switch the circuit on, rapidly at first but slowing down to a snail's pace at the end. Don't worry if your measurements of time and voltage aren't too accurate — we're only trying to prove the principle of the experiment, not the exact details. Besides, if you switch off

| Time (seconds) | Voltage (V) |
|:---:|:---|
| 5 | |
| 10 | |
| 15 | |
| 20 | |
| 30 | |
| 40 | |
| 50 | |
| 60 | |

**Table 4.1  Results of your measurements**

and then short circuit the capacitor, as already described, you can repeat the measurements as often as required.

When you've got your results, transfer them as points onto the blank graph of Figure 4.6 and sketch in a curve which goes approximately through all the points. Table 4.2 and Figure 4.7 show results obtained when the experiment was performed in preparing this book.

**Figure 4.6  A blank graph to plot your measurements. Use Table 4.1 above with it to record your measurements**

# Starting electronics

| Time (seconds) | Voltage (V) |
|:---:|:---:|
| 5 | 4 |
| 10 | 6 |
| 15 | 7 |
| 20 | 8 |
| 30 | 8.5 |
| 40 | 8.7 |
| 50 | 8.9 |
| 60 | 9 |

**Table 4.2    Our results while preparing this book**

**Figure 4.7    The results of our experiments**

What you should note in your experiment is that the voltage across the capacitor (any capacitor, in fact) rises, not instantaneously as with the voltage across a resistor, but gradually, following a curve. Is the curve the same, do you think, for all capacitors? Change the capacitor for one with a value of 220 $\mu$F (about half the previous one). Using the blank table of Table 4.3 and blank graph in Figure 4.8, perform the experiment again, to find out.

| Time (seconds) | Voltage (V) |
|:---:|:---|
| 5 | |
| 10 | |
| 15 | |
| 20 | |
| 25 | |
| 30 | |
| 40 | |

**Table 4.3    The results of your measurements**

**Figure 4.8    A blank graph to plot your measurements. Use Table 4.3 shown above as well**

Table 4.4 and Figure 4.9 show the results of our second experiment (yours should be similar) which shows that although the curves of the two circuits aren't exactly the same as far as the time axis is concerned, they are exactly the same shape. Interesting, huh?

This proves that the rising voltage across a capacitor follows some sort of law. It is an exponential law, and the curves you

# Starting electronics

| Time (seconds) | Voltage (V) |
|:---:|:---:|
| 5 | 6 |
| 10 | 7.2 |
| 15 | 8 |
| 20 | 8.3 |
| 25 | 8.5 |
| 30 | 8.7 |
| 40 | 9 |

**Table 4.4    The results of our second experiment**

**Figure 4.9    A graph showing our results of the second experiment**

obtained are known as exponential curves. These exponential curves are related by their time constants. We can calculate any rising exponential curve's time constant, as shown in Figure 4.10, where a value of 0.63 times the total voltage change cuts the time axis at a time equalling the curve's time constant, $\tau$ (the Greek letter tau).

In a capacitor circuit like that of Figure 4.4, the exponential curve's time constant is given simply by the product of the capacitor and resistor value. So, in the case of the first experiment, the time constant is:

$$\tau_1 \;=\; 470 \times 10^{-6} \times 15{,}000$$

$$=\; 7 \text{ seconds}$$

and, in the case of the second experiment:

$$\tau_2 \;=\; 220 \times 10^{-6} \times 15{,}000$$

$$=\; 3.3 \text{ seconds}$$

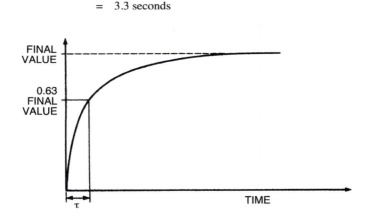

**Figure 4.10   A graph showing an exponential curve, related to the time constant**

You can check this, if you like, against your curves, or those in Figures 4.7 and 4.9, where you will find that the times when the capacitor voltage is about 5V7 (that is, 0.63 times the total voltage change of 9 V) are about 8 seconds and 4 seconds. Not bad when you consider possible experimental errors: the biggest of which is the existence of the meter resistance.

# Starting electronics

> **Hint:**
>
> One final point of interest about exponential curves, before we move on, is that the measurement denoted by the curve can be taken to be within about 1% of its final value after a time corresponding to five time constants. Because of this, it's taken as a rule-of-thumb by electronics engineers that the changing voltage across a capacitor is complete after $5\tau$.

## The other way

During these experiments you should have noticed how the voltage across the capacitor appeared to stay for a while, even after the switch had been turned off. How could this be so?

Figure 4.11 shows the circuit of an experiment you should now do, which will help you to understand this unusual capability of capacitors to apparently store voltage. Build the circuit up as in the breadboard layout of Figure 4.12, with the switch in its on position.

Now, measure the voltages at selected time intervals, and fill in the results into Table 4.5, then transfer them to form a curve in Figure 4.13, after switching the switch off.

90

**Figure 4.11    An experimental circuit to demonstrate how a capacitor stores voltage (or charge)**

**Figure 4.12    The breadboard layout for the experiment shown in Figure 4.11**

# Starting electronics

| Time (seconds) | Voltage (V) |
|:---:|:---:|
| 5 | |
| 10 | |
| 15 | |
| 20 | |
| 30 | |
| 40 | |
| 50 | |
| 60 | |

**Table 4.5   The results of your measurements**

**Figure 4.13   A blank graph to plot your measurements. Use Table 4.5 shown above**

The results we obtained in preparing this book are tabulated in Table 4.6 and plotted in Figure 4.14. You can see that the curve obtained is similar to the curves we got earlier, but are falling not rising. Like the earlier curves, this is an exponential curve, too.

92

| Time (seconds) | Voltage (V) |
|---|---|
| 5 | 7.5 |
| 10 | 6 |
| 15 | 5 |
| 20 | 4 |
| 30 | 2.5 |
| 40 | 1.5 |
| 50 | 1 |
| 60 | 0.7 |

**Table 4.6    The results of our experiments while preparing this book**

**Figure 4.14 The graph showing our own results**

If you wish, you can calculate the time constant here, in the same way, but remember that it must now be the time at which the voltage falls to 0.63 of the total voltage drop, that is, when:

$$V = 9 - (0.63 \times 9)$$

$$= 3.3V$$

93

## Theoretical aspects

So what is it about a capacitor which causes this delay in voltage between switching on or off the electricity supply and obtaining the final voltage across it? It's almost as if there's some mystical time delay inside the capacitor. To find the answer we'll have to consider a capacitor's innards again.

Figure 4.15 shows the capacitor we first saw in Figure 4.1, but this time it is shown connected to a battery which, as we know, is capable of supplying electrons from its negative terminal. A number of electrons gather on the capacitor plate connected to the battery's negative terminal, which in turn repel any electrons on the other plate towards the battery's positive terminal. A deficiency in electrons causes molecules of this capacitor plate to have a positive charge, so the two

**Figure 4.15** This is the capacitor shown in Figure 4.1, but this time connected to a battery with opposite charges built up on each plate

plates of the capacitor now have equal, but opposite, electric charges on them.

Current only flows during the time when the charges are building up on the capacitor plates — not before and not after. It's also important to remember that current cannot flow through the capacitor — a layer of insulator (known correctly as the dielectric) lies between the plates, remember — current only flows in the circuit around the capacitor.

If we now completely disconnect the charged capacitor from the battery, the equal and opposite charges remain — in theory — indefinitely. In practice, on the other hand, charge is always lost due to leakage current between the plates. You can try this for yourself, if you like: put a capacitor into the breadboard then charge it up by connecting the battery directly across it. Now disconnect the battery and leave the capacitor in the breadboard for a time (overnight, say). Then connect your meter across it to measure the voltage. You should still get a reading, but remember that the resistance of the meter itself will always drain the charge stored.

The size of the charge stored in a capacitor depends on two factors; the capacitor's capacitance (in farads) and the applied voltage. The relationship is given by:

$$Q = C \times V$$

where Q is the charge measured in coulombs, C is the capacitance and V is the voltage. From this we can see that a charge of one coulomb is stored by a capacitor of one farad, when a voltage of one volt is applied.

# Starting electronics

Alternatively, we may define the farad (as we promised we would, earlier in the chapter) as being the capacitance which will store a charge of one coulomb when a voltage of one volt is applied.

We can now understand why it is that changing the capacitor value changes the time constant, and hence changes the associated time delay in the changing voltage across the capacitor. Increasing the capacitance, say, increases the charge stored. As the current flowing is determined by the resistance in the circuit, and is thus fixed at any particular voltage, this increased charge takes longer to build up or longer to decay away. Reducing the capacitance reduces the charge, which is therefore more quickly stored or more quickly discharged.

Similarly, as the resistor in the circuit defines the current flowing to charge or discharge the capacitor, increasing or decreasing its value must decrease or increase the current, therefore increasing or decreasing the time taken to charge or discharge the capacitor. This is why the circuit's time constant is a product of both capacitance and resistance values.

## Capacitance values

Finally, we can look at how the size and construction of capacitors affects their capacitance. The capacitance of a basic capacitor for example, consisting of two parallel, flat plates separated by a dielectric may be calculated from the formula:

$$C = \varepsilon \times \frac{A}{d}$$

where $\varepsilon$ is the permittivity of the dielectric, A is the area of overlap of the plates, and d is the distance between the plates.

Different insulating materials have different permittivities, so different capacitor values may be made, simply by using different dielectrics. Air, say, has a permittivity of about $9 \times 10^{-12}$ farads per metre (that is, $9 \times 10^{-12} Fm^{-1}$). Permittivities of other materials are usually expressed as a relative permittivity, where a material's relative permittivity is the number of times its permittivity is greater than air. So, for example, mica (a typical capacitor dielectric), which has a relative permittivity of 6, has an actual permittivity of $54 \times 10^{-12} Fm^{-1}$ (that is, $6 \times 9 \times 10^{-12} Fm^{-1}$).

So, a possible capacitor, consisting of two parallel metal plates of overlapping area $20\,cm^2$, $10\,mm$ apart, with a mica dielectric, has a capacitance of:

$$C = 54 \times 10^{-12} \times \frac{0.002}{0.01}$$

$$= 10.8 \times 10^{-12}$$

$$= 10.8pF$$

Er, I think that's enough theory to be grappling with for now, don't you? Try your understanding of capacitors and how they work with the chapter's quiz.

# Starting electronics

## Quiz

Answers at end of book

1 The value of a capacitor with two plates of overlapping area $40\,cm^2$, separated by a 5 mm layer of air is:
a 7.2 pF
b $7.2 \times 10^{-12}pF$.
c 72 pF
d $100\,\mu F$
e a and b
f none of these

2 A capacitor of 1 nF has a voltage of 10 V applied across it, The maximum charge stored is:
a $1 \times 10^{-10}C$
b $1 \times 10^8C$
c $10 \times 10^{-8}C$
d $1 \times 10^{-8}F$
e a and c
f none of these

3 A capacitor of $10\,\mu F$ and a resistor of $1\,M\Omega$ are combined in a capacitor/resistor charging circuit. The time constant is:
a 1 second
b $10 \times 10^6$ second
c 10 seconds
d not possible to calculate from these figures
e none of these

4 A capacitor is fully charged by an applied voltage of 100 V, then discharged. After a time not less than 40 seconds the voltage across the capacitor has fallen to 1 V. At what time was the voltage across the capacitor 37 V?
a 10 seconds
b 37 seconds
c 8 seconds
d it is impossible to say from the figures given
e none of these

5 In question 4, the capacitor has a value of $100\,\mu F$. What is the charging resistance value?
a $80,000\,\mu F$
b 80,000 C
c $80\,k\Omega$
d $8\,k\Omega$
e a slap in the face with a wet fish
f none of these

# 5 ICs, oscillators and filters

You'll need a number of different components to build the circuits this chapter, some of them you'll already have, but there are a few new ones, too. First, the resistors required are:

● 1 x 1k5

● 1 x 4k7

● 1 x 10 k

Second, capacitors:

● 1 x 1 nF

● 2 x 10 nF

● 2 x 100 nF

- 1 x 1 $\mu$F electrolytic

- 1 x 10 $\mu$F electrolytic

Power ratings, tolerances and so on, of all these components are not critical, but the electrolytic capacitors should have a voltage rating of 9 V or more.

Some other components you already have: a switch, battery, battery connector, breadboard, multi-meter, should all be close at hand, as well as some single-strand tinned copper wire. You are going to use the single-strand wire to make connections from point-to-point on the breadboard, and as it's uninsulated this has to be done carefully, to prevent short circuits. Photo 5.1 shows the best method. Cut a short length of wire and hold it in the jaws of your snipe-nosed or long-nosed pliers. Bend the wire round the jaws to form a sharp right-angle in the wire. The tricky bit is next — judging the length of the connection you require, move the pliers along the wire and then bend the other end of the wire also at right angles round the other side of the jaws (Photo 5.2).

Photo 5.1    Take a short piece of wire between the jaws of your pliers...

**Photo 5.2**   ...and bend it carefully into two right angles, judging the length

If you remember that the grid of holes in the breadboard are equidistantly spaced at 2.5 mm (or a tenth of an inch if you're old-fashioned like me — what's an inch, Grandad?) then it becomes easier. A connection over two holes is 5 mm long, over four holes is 10 mm long and so on — you'll soon get the hang of it.

Now, holding the wire at the top, in the pliers, push it into the breadboard as shown in Photo 5.3, until it lies flush on the surface of the breadboard as in Photo 5.4. No bother, eh? Even with a number of components in the breadboard it is difficult to short circuit connections made this way.

Two other components you need are:

● 1 x 555 integrated circuit

● 1 x light emitting diode (any colour).

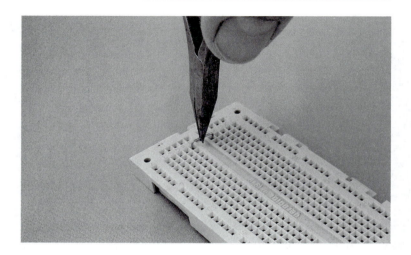

Photo 5.3    The wire link you have made should drop neatly into the breadboard

Photo 5.4    If the link is a good fit. It can be used over and over again in different positions

# ICs oscillators and filters

We've seen an integrated circuit (IC) before and we know what it looks like, but we've never used one before, so we'll take a closer look at the 555 now. It is an 8-pin dual-in-line (DIL) device and one is shown in Photo 5.5. Somewhere on its body is a notch or dot, which indicates the whereabouts of pin 1 of the IC as shown in Figure 5.1. The remainder of the pins are numbered in sequence in an anti-clockwise direction around the IC.

Photo 5.5   The 555, an 8 pin DIL timer IC, beside a UK one penny piece

Figure 5.1   A diagram of the normal configuration of any IC: the dot marks pin 1 and the remaining pins run anti-clockwise

ICs should be inserted into a breadboard across the bread-board's central bridged portion. Isn't it amazing that this portion is 7.5 mm across and, hey presto, the rows of pins of the IC are about 7.5 mm apart? It's as if the IC was made for the breadboard! So the IC fits into the breadboard something like that in Photo 5.6.

Photo 5.6   The IC mounted across the central divide in the breadboard, which is designed to be the exact size

We'll look at the light emitting diode later.

If you remember, last chapter we took a close look at capaci-tors, how they charge and discharge, storing and releasing electrical energy. The first thing we shall do this chapter, however, is use this principle to build a useful circuit called an oscillator. Then, in turn, we shall use the oscillator to show some more principles of capacitors. So, we've got a two-fold job to do now and there's an awful lot of work to get through — let's get started.

Figure 5.2 shows the circuit of the oscillator we're going to build. It's a common type of oscillator known as an astable multi-vibrator. The name arises because the output signal appears to oscillate (or vibrate) between two voltages, never resting at one voltage for more than just a short period of time (it is therefore unstable, more commonly known as astable). An astable multi-vibrator built from discrete that is, individual components, can be tricky to construct so we've opted to use an integrated circuit (the 555) as the oscillator's heart.

**Figure 5.2    The circuit of the astable multi-vibrator circuit using the 555**

Inside the 555 is an electronic switch which turns on when the voltage across it is approximately two-thirds the power supply voltage (about 6 V in the circuit of Figure 5.2), and off when the voltage is less than one-third the power supply voltage (about 3 V). Figure 5.3(a) shows an equivalent circuit to that of Figure 5.2 for the times during which the electronic switch of the 555 is off.

You should be able to work out that the capacitor C1 of the circuit is connected through resistors R1 and R2 to the positive power supply rail. The time constant $\tau_1$ of this part of the circuit is therefore given by:

$$\tau_1 = RC = (R1 + R2)C$$

When the voltage across the switch rises to about 6 V, however, the switch turns on (as shown in Figure 5.3(b), forming a short circuit across the capacitor and resistor, R2. The capacitor now discharges with a time constant given by:

$$\tau_2 = R2 \times C$$

Of course, when the discharging voltage across the switch falls to about 3 V, the switch turns off again, and the capacitor charges up once more. The process repeats indefinitely, with the switch turning on and off at a rate determined by the two time constants. Because of this up and down effect such oscillators are often known as relaxation oscillators.

As you might expect the circuit integrated within the 555 is not just that simple and there are many other parts to it (one part, for example, converts the charging and discharging exponential voltages into only two definite voltages — 9 V and 0 V — so that the 555's output signal is a square wave, as shown in Figure 5.3(c). But the basic idea of the astable multi-vibrator formed by a 555 is just as we've described here.

**Figure 5.3** (a) shows a circuit equivalent to Figure 5.2 when the 555's switch is off; (b) shows the circuit when the switch is on; (c) shows the square wave output signal

## Throwing light on it

The 555 IC is one of only two new types of electronic component this circuit introduces you to. The other is a light emitting diode (LED) which is a type of indicator. One is shown in Photo 5.7. LEDs are polarised and so must be inserted into circuit the right way round. All LEDs have an anode (which goes to the more positive side of the circuit) and cathode (which goes to the more negative side).

Generally, but not always, the anode and cathode of an LED are identified by the lengths of the component leads — the cathode is the shorter of the two. There's often a flat side to an LED too — usually, but again not always, on the cathode side — to help identify leads.

107

# Starting electronics

**Photo 5.7**   One of the most popular electronic components, the LED

The complete circuit's breadboard layout is shown in Figure 5.4. Build it and see what happens.

When you turn on, you should find that the LED flashes on and off, quite rapidly (about five or six times a second, actually). This means your circuit is working correctly. If it doesn't work check polarity of all polarised components: the battery, IC, LED and capacitor.

**Figure 5.4**   The breadboard layout for the multi-vibrator circuit as shown in Figure 5.2

## Ouch, that hertz

We can calculate the rate at which the LED flashes, more accurately, from formulae relating to the 555. A quick study of the square wave output shows that it consists of a higher voltage for a time (which we can call T1) and a lower voltage for a time (which we will call T2).

Now, T1 is given by:

$$T1 = 0.7\,\tau_1$$

and T2 is given by:

$$T2 = 0.7\,\tau_2$$

So, the time for the whole period of the square wave is:

$$T1 + T2 = 0.7(\tau_1 + \tau_2)$$

and as the frequency of a waveform is the inverse of its period we may calculate the waveform's frequency as:

$$f = \frac{1}{0.7(\tau_1 + \tau_2)}$$

Earlier, we defined the two time constants, $\tau_1$ and $\tau_2$, as functions of the capacitor and the two resistors, and so by substituting them into the above formula, we can calculate the frequency as:

$$f = \frac{1}{0.7C(R1 + 2R2)} \tag{1}$$

So, the frequency of the output signal of the circuit of Figure 5.2 is:

$$f = \frac{1}{0.7 \times 10 \times 10^{-6} \times (4700 + 20{,}000)}$$

$$= 5.8 \text{ cycles per second}$$

or, more correctly speaking:

$$= 5.8 \text{ Hertz (shortened to 5.8Hz)}$$

Equation 1 is quite important really, because it shows that the frequency of the signal is inversely proportional to the capacitance. If we decrease the value of the capacitor we will increase the frequency. We can test this by taking out the $10\,\mu$F capacitor and putting in a $1\,\mu$F capacitor. Now, the LED flashes so quickly (about 58 times a second) that your eye can't even detect it is flashing and it appears to be always on. If you replace the capacitor with one of a value of say $100\,\mu$F the LED will flash only very slowly.

Now, let's stop and think about what we've just done. Basically we've used a capacitor in precisely the ways we looked at last chapter — to charge and discharge with electrical energy so that the voltage across the capacitor goes up and down at the same time. True, in the experiments last chapter you were the switch, whereas this chapter an IC has taken your place. But the principle — charging and discharging a capacitor — is the same.

The current which enters the capacitor to charge it, then leaves the capacitor to discharge it, is direct current because it comes from a 9 V d.c. battery. However, if we look at the output signal (Figure 5.3(c)) we can see that the signal alternates between two levels. Looked at in this way, the astable multi-vibrator is a d.c.-to-a.c.converter. And that is going to be useful in our next experiment, where we look at the way capacitors are affected by a.c. The circuit we shall look at is shown in Figure 5.5 and is very simple, but it'll do nicely, thank you. It should remind you of a similar circuit we have already looked at; the voltage divider, only one of the two resistors of the voltage divider has been replaced by a capacitor. Like an ordinary voltage divider the circuit has an input and an output. What we're going to attempt to do in the experiment is to measure the output signal when the input signal is supplied from our astable multivibrator.

**Figure 5.5    A voltage divider – but with a capacitor**

Figure 5.6 shows the whole circuit of the experiment while Figure 5.7 shows the breadboard layout. The procedure for the experiment is pretty straightforward: measure the output voltage of the a.c. voltage divider when a number of different frequencies are generated by the astable multi-vibrator, then tabulate and plot these results on a graph. Things really couldn't be easier. Table 5.1 is the table to fill in as you obtain your results and Figure 5.8 is marked out in a suitable grid to plot your graph. To change the astable multivibrator's frequency, it is only necessary to change capacitor Cl. Increasing it ten-fold decreases the frequency by a factor of

# Starting electronics

**Figure 5.6    A circuit combining the astable multi-vibrator and the circuit in Figure 5.5**

**Figure 5.7    The breadboard layout of the circuit in Figure 5.6. It is the same as that in Figure 5.4, with a capacitor in place of the LED**

| Value of C1 | Calculated frequency | Measured voltage |
|---|---|---|
| 10 µF | 5.8 Hz | |
| 1 µF | 58 Hz | |
| 100 nF | 580 Hz | |
| 10 nF | 5.8 kHz | |
| 1 nF | 58 kHz | |

**Table 5.1    Results when capacitor C2 is 100 nF**

**Figure 5.8    A blank graph on which you can plot your experimental results. Use Table 5.1 above as well**

ten; decreasing the capacitor value by ten increases the frequency ten-fold. Five different values of capacitor therefore give an adequate range of frequencies.

As you do the experiment you'll find that only quite low voltages are measured (up to about 4 V a.c.) so depending on your

# Starting electronics

multi-meter's lowest a.c. range you may not achieve the level of accuracy you would normally desire, but the results will be OK, nevertheless.

Table 5.2 and Figure 5.9 show my results which should be similar to yours (if not, you're wrong — I can't be wrong, can I?). The graph shows that the size of the output signal of the a.c.

| Value of C1 | Calculated Frequency | Measured voltage |
|:---:|:---:|:---:|
| 10 µF | 5.8Hz | 0V |
| 1 µF | 58Hz | 0V |
| 100 nF | 580Hz | 1.5V |
| 10 nF | 5.8 kHz | 4V |
| 1 nF | 58 kHz | 4.2V |

**Table 5.2   My tabled results**

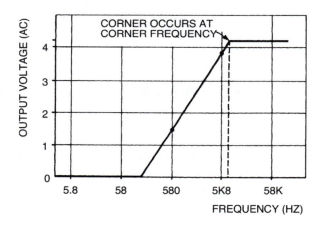

**Figure 5.9   Graphed results of Table 5.2**

voltage divider is dependent on the frequency of the applied input signal. In particular, there are three clearly distinguishable sections to this graph, each relating to frequency.

First, above a certain frequency, known as the corner frequency, the output signal is constant and at its maximum.

Second, at low frequencies (close to 0 Hz) the output is zero.

Third, between these two sections the output signal varies in size depending on the applied input signal frequency.

Is this the same for all a.c. voltage dividers of the type shown in Figure 5.5? Well, let's repeat the experiment using a different capacitor for C2, to find out.

Try a 10 nF capacitor and repeat the whole procedure, putting your results in Table 5.3 and Figure 5.10. Table 5.4 and Figure 5.11 show our results.

| Value of C1 | Calculated frequency | Measured voltage |
|---|---|---|
| 10 μF | 5.8 Hz | |
| 1 μF | 58 Hz | |
| 100 nF | 580 Hz | |
| 10 nF | 5.8 kHz | |
| 1 nF | 58 kHz | |

**Table 5.3    Results when capacitor C2 is 10 nF**

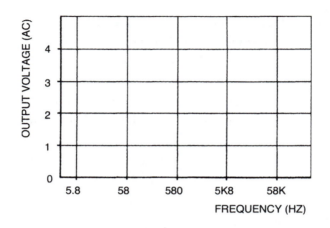

**Figure 5.10    A graph for use with your results from the experiment with the 10 nF capacitor in Table 5.3**

And yes, the graph is the same shape but is moved along the horizontal axis by an amount equivalent to a ten-fold increase in frequency (the capacitor was decreased in value by ten-fold, remember). A similar inverse relationship is caused by changing the resistor value, too.

Frequency, capacitance and resistance are related in the a.c. voltage divider by the expression:

$$f = \frac{1}{RC}$$

where f is the corner frequency. For the first voltage divider, with a capacitance of 10 nF and a resistance of 1.5 kΩ, the corner frequency is:

$$f = \frac{1}{1500 \times 100 \times 10^{-9}}$$

$$= 6666\text{Hz}$$

116

| Value of C1 | Calculated frequency | Measured voltage |
|---|---|---|
| 10 μF | 5.8 Hz | 0V |
| 1 μF | 58 Hz | 0V |
| 100 nF | 580 Hz | 0V |
| 10 nF | 5.8 kHz | 1.5V |
| 1 nF | 58 kHz | 4V |

**Table 5.4   My results (C2 = 10 nF)**

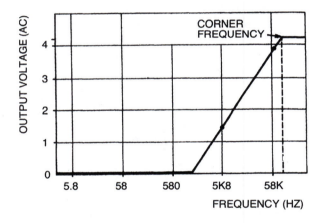

**Figure 5.11   Graph of my results from Table 5.4 in the same experiment**

which is more or less what we found in the experiment. In the second voltage divider, with a 10 nF capacitor, the corner frequency increases by ten to 66,666 Hz.

Remembering what we learned last chapter about resistors and capacitors in charging/discharging circuits, we can simplify the expression for corner frequency to:

117

$$f = \frac{1}{\tau}$$

because the product RC is the time constant, $\tau$. This may be easier for you to remember.

An a.c. voltage divider can be constructed in a different way, as shown in Figure 5.12. Here the resistor and capacitor are transposed. What do you think the result will be? Well, the output signal size now decreases with increasing frequency — exactly the opposite effect of the a.c. voltage divider of

**Figure 5.12    An a.c. voltage divider with the resistor and capacitor transposed**

Figure 5.5! All other aspects are the same, however: there is a constant section below a corner frequency, and a section where the output signal is zero, as shown in Figure 5.13. Once again the corner frequency is given by the expression:

$$f = \frac{1}{RC} = \frac{1}{\tau}$$

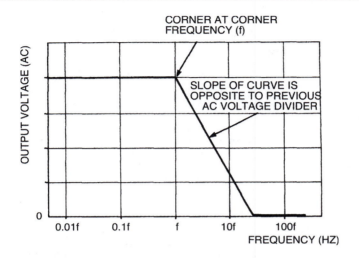

**Figure 5.13   This graph shows the similarities and differences between this circuit and that in Figure 5.5**

## Filter Tips

The a.c. voltage dividers of Figures 5.5 and 5.12 are normally shown in a slightly different way, as in Figures 5.14(a) and (b). Due to the fact they allow signals of some frequencies to pass through, while filtering out other signal frequencies, they are more commonly called filters.

The filter of Figure 5.14(a) is known as a high-pass filter — because it allows signal frequencies higher than its corner frequency to pass while filtering out signal frequencies lower than its corner frequency.

119

# Starting electronics

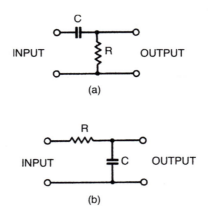

(a)

(b)

**Figure 5.14    The a.c. divider sections of the circuits in Figure 5.5 and Figure 5.12**

The filter of Figure 5.14(b) is a low-pass filter — yes, you've guessed it — because it passes signals with frequencies below its corner frequency, while filtering out higher frequency signals.

Filters are quite useful in a number of areas of electronics. The most obvious example of a low-pass filter is probably the scratch filter sometimes seen on stereo systems. Scratches and surface noise when a record is played, or tape hiss when a cassette tape is played, consist of quite high frequencies; the scratch filter merely filters out these frequencies, leaving the music relatively noise free.

Bass and treble controls of an amplifier are also examples of high- and low-pass filters: a bit more complex than the

120

simple ones we've looked at here but following the same general principles.

And that's about it for this chapter. You can try a few experiments of your own with filters if you want. Just remember that whenever you use your meter to measure voltage across a resistor in a filter, the meter resistance affects the actual value of resistance and can thus drastically affect the reading.

Now try the quiz over the page to check and see if you've been able to understand what we've been looking at this chapter.

# Starting electronics

## Quiz

Answers at end of book

1 A signal of frequency 1 kHz is applied to a low-pass filter with a corner frequency of 10 kHz. What happens?

a The output signal is one-tenth of the input signal
b The output signal is larger than the input signal
c There is no output signal
d The output signal is identical to the input signal
e All of these

2 A high-pass filter consisting of a 10 kΩ resistor and an unknown capacitor has a corner frequency of 100 Hz. What is the value of the capacitor?

a 1 nF
b 10 nF
c 100 nF
d 1000 nF
e 1 μF

3 A low-pass filter consisting of a 1 μF capacitor and a unknown resistor has a corner frequency of 100 Hz. What is the value of the resistor?

a 10 kΩ
b 1 kΩ
c 100 kΩ
d All of these
e It makes no difference
f d and e
g None of these

4 In a circuit similar to that in Figure 5.2, resistor R1 is 10 kΩ, resistor R2 is 100 kΩ, and capacitor C1 is 10 nF. The output frequency of the astable multi-vibrator is:

a a square wave
b about 680 Hz
c too fast to see the LED flashing
d a and b
e c and d
f None of these

# 6    Diodes I

We're going to take a close look at a new type of component this chapter — the diode. Diodes are the simplest component in the range of devices known as semiconductors. Actually, we've briefly looked at a form of diode before: the light-emitting diode, or LED, and we've seen another semiconductor too: the 555 integrated circuit. But now we'll start to consider semiconductors in depth.

Naturally, you'll need some new components for the circuits you're going to build here. These are:

● 1 x 150 Ω, 0.5 W resistor

● 1 x 1N4001 diode

● 1 x OA47 diode

● 1 x 3V0 zener diode (type BZY88)

● 1 x 1k0 miniature horizontal preset

# Starting electronics

Diodes get their name from the basic fact that they have two electrodes (di — ode, geddit?). One of these electrodes is known as the anode: the other is the cathode. Figure 6.1 shows the symbol for a diode, where the anode and cathode are marked. Figure 6.2 shows some typical diode body shapes, again with anode and cathode marked.

Photo 6.1 is a photograph of a miniature horizontal preset resistor. We're going to use it in the following circuits as a variable voltage divider. To adjust it you'll need a small screwdriver or tool to fit in the adjusting slot — turning it one way and another alters position of the preset's wiper over the resistance track.

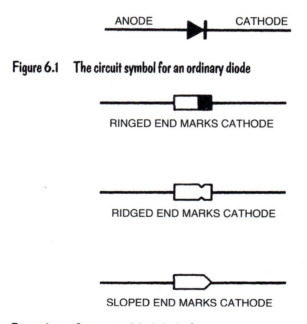

**Figure 6.1   The circuit symbol for an ordinary diode**

RINGED END MARKS CATHODE

RIDGED END MARKS CATHODE

SLOPED END MARKS CATHODE

**Figure 6.2   Some typical diode body shapes**

124

**Photo 6.1   A horizontal preset resistor**

Figure 6.3  shows the circuit we're going to build first this chapter. It's very simple, using two components we're already familiar with (a resistor and an LED) together with the new component we want to look at: a diode. Before you build it, note which way round the diode is and also make sure you get the LED polarised correctly, too. In effect, the anodes of each diode (a LED is a diode, too, remember — a light emit-

**Figure 6.3   Our first simple circuit using a diode**

125

# Starting electronics

ting diode) connect to the more positive side in the circuit. A breadboard layout is shown in Figure 6.4, though by this stage you should perhaps be confidently planning your own breadboard layouts.

**Figure 6.4   A breadboard layout for the circuit in Figure 6.3**

## Which way round?

If you've connected the circuit up correctly, the LED should now be on. This proves that current is flowing. To calculate exactly what current we can use Ohm's law. Let's assume that the total battery voltage of 9 V is dropped across the resistor and that no voltage occurs across the two diodes. In fact, there is voltage across the diodes, but we needn't worry about it yet, as it is only a small amount. We'll measure it, however, soon.

126

Now, with a resistance of 1k5, and a voltage of 9 V, we can calculate the current flowing, as:

$$I = \frac{V}{R} = \frac{9}{1500} = 6mA$$

The next thing to do is to turn around the diode, so that its cathode is more positive, as in the circuit of Figure 6.5. The breadboard layout is the same with anode and cathode reversed, so we needn't redraw it.

**Figure 6.5    The circuit again, but with the diode reversed**

What happens? You should find that absolutely nothing happens. The LED does not light up, so no current must be flowing. The action of reversing the diode has resulted in the stopping of current. We can summarise this quite simply in Figure 6.6.

Figure 6.6(a) shows a diode whose anode is positive with respect to its cathode. Although we've shown the anode as positive with a + symbol, and the cathode as negative with

127

a – symbol, they don't necessarily have to be positive and negative. The cathode could for example be at a voltage of +1000 V if the anode was at a greater positive voltage of, say +1001 V. All that needs to occur is that the anode is positive with respect to the cathode.

Under such a condition, the diode is said to be forward biased and current will flow, from anode to cathode.

When a diode is reverse biased i.e., its cathode is positive with respect to the anode, no current flows, as shown in Figure 6.6(b). Obviously, something happens within the diode which we can't see, depending on the polarity of the applied voltage to define whether current can flow or not. Just exactly what this something is, isn't necessary to understand

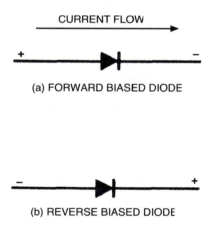

Figure 6.6    Circuit diagrams for forward and reverse biased diodes

here. We needn't know any more about it here because we're only concerned with the practical aspects at the moment; and all we need to remember is that a forward biased diode conducts, allowing current to flow, while a reverse biased diode doesn't.

What we do need to consider in more detail; however, is the value of the current flowing, and the small, but nevertheless apparent, voltage which occurs across the diode, when a diode is forward biased (the voltage we said earlier we needn't then worry about). The following experiment will show how the current and the voltage are related.

Figure 6.7 shows the circuit you have to build. You'll see that two basic measurements need to be taken with your meter. The first measurement is the voltage across the forward biased diode, the second measurement is the current through it. Each measurement needs to be taken a number of times as the preset is varied in an organised way. Table 6.1, which is half complete, is for you to record your results, and Figure 6.8 is a blank graph for you to plot the results into a curve. Do the experiment the following way:

**Figure 6.7    A circuit to test the operation of a forward biased diode**

| Current (mA) | Voltage (V) |
|---|---|
| 0 | 0 |
| | 0.4 |
| | 0.6 |
| 2 | |
| 5 | |
| 10 | |
| 20 | |

**Table 6.1** This is half complete, add the results of your experiment

**Figure 6.8** A blank graph for you to plot the results of your experiment

1)     set up the components on the breadboard to measure only the voltage across the diode. The breadboard layout is given in Figure 6.9. Before you connect your battery to the circuit, make sure the wiper of the preset is turned fully anti-clockwise,

**Figure 6.9    The breadboard layout for the circuit in Figure 6.7**

2)      adjust the preset wiper clockwise, until the first voltage in Table 6.1 is reached,

3)      now set up the breadboard layout of Figure 6.10, to measure the current through the diode — the breadboard layout is designed so that all you have to do is take out a short length of single-strand connecting wire and change the position of the meter and its range. Record the value of the current at the voltage of step 2,

4)      change the position of the meter and its range, and replace the link in the breadboard so that voltage across the diode is measured again,

5)      repeat steps 2, 3 and 4 with the next voltage in the table,

**Figure 6.10   The same circuit, set up to measure the current through the diode**

6)      repeat step 5 until the table shows a given current reading. Now set the current through the diode to this given value and measure and record the voltage,

7)      set the current to each value given in the table and record the corresponding voltage,  until the table is complete.

## Tricky

In this way, first measuring voltage then measuring current, or first measuring current then voltage, changing the position and range of the meter, as well as removing or inserting the link depending on whether you're measuring current or voltage, the experiment can be undertaken. Yes, it's tricky, but we never said it was a doddle, did we? You'll soon get the hang of it and get some good results.

Now plot your results on the graph of Figure 6.8. My results (correct naturally!) are shown in Table 6.2 and Figure 6.11.

| Current (mA) | Voltage (V) |
|:---:|:---:|
| 0 | 0 |
| 0 | 0.4 |
| 1 | 0.6 |
| 2 | 0.65 |
| 5 | 0.65 |
| 10 | 0.7 |
| 20 | 0.75 |

**Table 6.2    The results of my own experiment**

**Figure 6.11    My own results from the experiment**

# Starting electronics

Repeat the whole experiment again, using the 0A47 diode, this time. You can put your results in Table 6.3 and plot your graph in Figure 6.12. Our results are in Table 6.4 and Figure 6.13.

| Current (mA) | Voltage (V) |
|---|---|
| 0 | 0 |
| | 0.1 |
| | 0.2 |
| | 0.25 |
| | 0.3 |
| 3 | |
| 5 | |
| 10 | |
| 20 | |

**Table 6.3** The results of your experiment

**Figure 6.12** Use this graph to plot your results from the second experiment

134

| Current (mA) | Voltage (V) |
|:---:|:---:|
| 0 | 0 |
| 0 | 0.1 |
| 0.5 | 0.2 |
| 1 | 0.25 |
| 2 | 0.3 |
| 3 | 0.32 |
| 5 | 0.35 |
| 10 | 0.4 |
| 20 | 0.45 |

**Table 6.4  My results from the second experiment**

**Figure 6.13  A graph plotting my results from the second experiment**

As you might expect, these two plotted curves are the same basic shape. The only real difference between them is that they change from a level to an extremely steep line at different positions. The OA47 curve, for example, changes at about 0V3, while the 1N4001 curve changes at about 0V65.

The sharp changes in the curves correspond to what are sometimes called transition voltages — the transition voltage for the OA47 is about 0V3, the transition voltage for the 1N4001 is about 0V65. It's important to remember, though, that the transition voltages in these curves are only for the particular current range under consideration — 0 to 20 mA in this case. If similar curves are plotted for different current ranges then slightly different transition voltages will be obtained. In any current range, however, the transition voltages won't be more than about 0.V1 different to the transition voltages we've seen here. The two curves — of the OA47 and the 1N4001 diodes — show that a different transition voltage is obtained (0V3 for the OA47, 0V65 for the 1N4001) depending on which semiconductor material a diode is made from. The OA47 diode is made from germanium while the 1N4001 is of silicon construction. All germanium diodes have a transition voltage of about 0V2 to 0V3; similarly all silicon diodes have a transition voltage of about 0V6 to 0V7.

These two curves are exponential curves — in the same way that capacitor charge/discharge curves (see Chapter 4) are exponential, just in a different direction, that's all — and form part of what are called diode characteristic curves or sometimes simply diode characteristics. But the characteristics we have determined here are really only half the story as far as diodes are concerned. All we have plotted are the forward voltages and resultant forward currents when the diodes are forward biased. If diodes were perfect this would be all the information we need. But, yes you've guessed it, diodes are not perfect — when they are reverse biased so that they have reverse voltages, reverse currents flow. So, to get a true picture of diode operation we have to extend the characteristic curves to include reverse biased conditions.

Reverse biasing a diode means that its anode is more negative with respect to its cathode. So by interpolating the x- and y-axes of the graph, we can provide a grid from the diode characteristic which allows it to be drawn in both forward and reverse biased conditions.

It wouldn't be possible for you to plot the reverse biased conditions, for an ordinary diode, the way you did the forward biased experiment (we will, however, do it for a special type of diode soon), so instead we'll make it easy for you and give you the whole characteristic curve. Whatever type of diode, it will follow a similar curve to that of Figure 6.14, where the important points are marked.

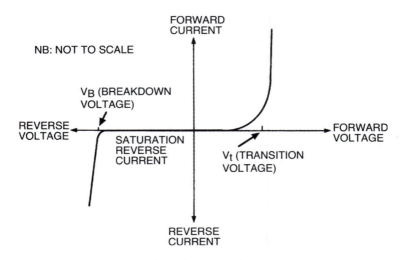

Figure 6.14    Plotting the reverse bias characteristics for an ordinary diode is not practical, so we give you the characteristic curve here

## Reverse bias

From the curve you can see that there are two distinct parts which occur when a diode is reverse biased. First, at quite low reverse voltages, from about –0.V1 to the breakdown voltage there is a more or less constant but small reverse current. The actual value of this reverse current (known as the saturation reverse current, or just the saturation current) depends on the individual diode, but is generally in the order of microamps.

The second distinct part of the reverse biased characteristic occurs when the reverse voltage is above the breakdown voltage. The reverse current increases sharply with only comparatively small increases in reverse voltage. The reason for this is because of electronics breakdown of the diode when electrons gain so much energy due to the voltage that they push into one another just like rocks and boulders rolling down a steep mountainside push into other rocks and boulders which, in turn, start to roll down the mountainside pushing into more rocks and boulders, forming an avalanche. This analogy turns out to be an apt one, and in fact the electronic diode breakdown voltage is sometimes referred to as avalanche breakdown and the breakdown voltage is sometimes called the avalanche voltage. Similarly the sharp knee in the curve at the breakdown voltage is often called the avalanche point.

In most ordinary diodes the breakdown voltage is quite high (in the 1N4001 it is well over –50 V), so this is one reason why you couldn't plot the whole characteristic curve, including reverse biased conditions, in the same experiment — our battery voltage of 9 V simply isn't high enough to cause breakdown.

Some special diodes, on the other hand, are purposefully manufactured to have a low breakdown voltage, and you can use one of them to study and plot your own complete diode characteristic. Such diodes are named after the American scientist, Zener, who was one of the first people to study electronic breakdown. The zener diode you are going to use is rated at 3V0 which of course means that its breakdown occurs at 3 V which is below the battery voltage and is therefore plottable in the same sort of experiment as the last one. The symbol for a zener diode, incidentally, is shown in Figure 6.15.

CATHODE        ANODE

**Figure 6.15**    The circuit symbol for a zener diode

**Hint:**

The way to remember the zener diode circuit symbol is to note that the bent line representing its cathode corresponds to the electronics breakdown.

Procedure is more or less the same as before. The circuit is shown in Figure 6.16 where the zener diode is shown forward biased. Complete Table 6.5 with the circuit as shown, then turn

**Figure 6.16  A circuit with the zener diode forward biased**

| Current (mA) | Voltage (V) |
|---|---|
| | 0 |
| | 0.4 |
| | 0.6 |
| 1 | |
| 2 | |
| 9 | |
| 5 | |
| 10 | |
| 20 | |

**Table 6.5  Show the results of your experiment**

the zener diode round as shown in the circuit of Figure 6.17 (this is the way zener diodes are normally used) so that it is reverse biased, then perform the experiment again completing Table 6.6 as you go along. Although all the results will in theory be negative, you don't need to turn the meter round or anything — the diode has been turned around remember, and so is already reverse biased.

140

**Figure 6.17   The circuit, with the zener diode reverse biased**

| -Current (mA) | -Voltage (V) |
|:---:|:---:|
| 0 | 0 |
| | 1 |
| | 2 |
| | 2.5 |
| | 3 |
| 10 | |
| 20 | |

**Table 6.6   To show the results of your further experiments**

Next, plot your complete characteristic on the graph of Figure 6.18. My results and characteristics are in Table 6.7 and 6.8, and Figure 6.19. Yours should be similar.

From the zener diode characteristic you will see that it acts like any ordinary diode. When forward biased it has an exponential curve with a transition voltage of about 0V7 for the current range observed. When reverse biased, on the other hand, you can see the breakdown voltage of about –3 V which

141

# Starting electronics

FORWARD CURRENT (mA)

REVERSE VOLTAGE (V)

FORWARD VOLTAGE (V)

REVERSE CURRENT (mA)

**Figure 6.18    Plot your results from the zener diode experiment on this graph. Also use Table 6.6 shown on previous page**

| Current (mA) | Voltage (V) |
|---|---|
| 0 | 0 |
| 0 | 0.4 |
| 0 | 0.6 |
| 1 | 0.7 |
| 2 | 0.75 |
| 5 | 0.75 |
| 10 | 0.8 |
| 20 | 0.8 |

**Table 6.7    The results of my experiments**

| -Current (mA) | -Voltage (V) |
|---|---|
| 0 | 0 |
| 0 | 1 |
| 0.5 | 2 |
| 2 | 2.5 |
| 5 | 3 |
| 10 | 3.2 |
| 20 | 3.5 |

**Table 6.8    More results from my experiments**

142

FORWARD CURRENT (mA)

REVERSE VOLTAGE (V)

FORWARD VOLTAGE (V)

REVERSE CURRENT (mA)

**Figure 6.19    My results from the zener diode experiment are given in this graph and in Tables 6.7 and 6.8**

occurs when the reverse current appears to be zero, but in fact, a small saturation current does exist — it was simply too small to measure on the meter.

Voila — a complete diode characteristic curve.

Finally, it stands to reason that any diode must have maximum ratings above which the heat generated by the voltage and current is too much for the diode to withstand. Under such circumstances the diode body may melt (if it is a glass diode such as the OA47), or, more likely, it will crack and fall apart. To make sure their diodes don't encounter such rough treatment manufacturers supply maximum ratings which should not be exceeded. Typical maximum ratings of the two ordinary diodes we have looked at; the 1N4001 and the 0A47 are listed in Table 6.9.

| Maximum ratings | 1N4001 | OA47 |
|---|---|---|
| Maximum mean forward current | 1 A | 110 mA |
| Maximum repetitive forward current | 10 A | 150 mA |
| Maximum reverse voltage | –50 V | –25 V |
| Maximum operating temperature | 170°C | 60°C |

**Table 6.9  Typical maximum ratings of the two diodes we have been looking at**

Next chapter we shall be considering diodes again, and how we can use them, practically, in circuits; what their main uses are, and how to choose the best one for any specific purpose.

# 7   Diodes I

In the last chapter, we took a detailed look at diodes and their characteristic curves. This chapter we're going to take this one stage further and consider how we use the characteristic curve to define how the diode will operate in any particular circuit. Finally, we'll look at a number of circuits which show some of the many uses of ordinary diodes.

The components you'll need for the circuits this chapter are:

> 1 x 10 k resistor
>
> 2 x 15 k resistors
>
> 2 x 100 k resistors
>
> 1 x 10 $\mu$F 10 V electrolytic capacitor

# Starting electronics

1 x 220 μF 10 V electrolytic capacitor

1 x 1N4001 diode

1 x LED

1 x 555 integrated circuit

Figure 7.1 shows the forward biased section of a typical diode characteristic curve. It has a transition voltage of about 0V7, so you should know that it's the characteristic curve of a silicon diode.

Diodes aren't the only electronic components for which characteristic curves may be drawn — most components can be studied in this way. After all, the curve is merely a graph of the voltage across the component compared to the current through it. So, it's equally possible that we draw a characteristic curve of, say, a resistor. To do this we could perform

**Figure 7.1    The forward biased section of the characteristic curve of a silicon diode**

146

the same experiment we did last chapter with the diodes: measuring the voltage and current at a number of points, then sketching the curve as being the line which connects the points marked on the graph.

But there's no need to do this in the case of a resistor, because we know that resistors follow Ohm's law. We know that:

$$R = \frac{V}{I}$$

where R is the resistance, V is the voltage across the resistor, and I is the current through it. So, for any value of resistor, we can choose a value for, say, the voltage across it, and hence calculate the current through it. Figure 7.2 is a blank graph. Calculate and then draw on the graph characteristic curves for two resistors: of values $100\,\Omega$ and $200\,\Omega$. The procedure is simple: just calculate the current at each voltage point for each resistor.

Figure 7.2    A blank graph for you to fill in — see the text above for instructions

Your resultant characteristic curves should look like those in Figure 7.3 — two straight lines.

Figure 7.3   Your efforts with Figure 7.2 should produce a graph something like this

Take note — Take note — Take note — Take note

Resistor characteristic curves are straight lines because resistors follow Ohm's law (we say they are ohmic) and Ohm's law is a linear relationship. So resistor characteristic curves are linear, too. And because they are linear there's really no point in drawing them, and you'll never see them in this form anywhere else — we drew them simply to emphasise the principle.

148

# Diodes II

We need to draw diode characteristic curves, on the other hand, because they're non-ohmic and hence, non-linear. So to see what current passes through the device with any particular voltage across it, it's useful to see its characteristic curve. This is generally true of any semiconductor device, as we'll see in later chapters.

## Maths

Although diodes are non-ohmic, this doesn't mean that their operation can't be explained mathematically (just as Ohm's law or, V = IR, say is a mathematical formula). Diodes, in fact, follow a relationship every bit as mathematical as Ohm's law. The relationship is:

$$I = I_s \left\{ \frac{e(qV)}{kT} - 1 \right\}$$

where I is the current through the diode, $I_s$ is the saturation reverse current, q is the magnitude of an electron's charge, k is Boltzmann's constant, and T is the absolute temperature in degrees Kelvin. As q and k are both constant and at room temperature the absolute temperature is more or less constant, the part of the equation q/kT is also more or less constant at about 40 (work it out yourself if you want: q is $1.6 \times 10^{-19}$ C; k is $1.38 \times 10^{-23}$ JK$^{-1}$ and room temperature, say, 17°C is 290 K.

The equation is thus simplified to be approximately:

$$I = I_s \left( e^{40V} - 1 \right)$$

The exponential factor ($e^{40V}$), of course confirms what we already knew to be true — that the diode characteristic curve is an exponential curve. From this characteristic equation we may calculate the current flowing through a diode for any given voltage across it, just as the formulae associated with Ohm's law do the same for resistors.

> **Hint:**
>
> _____
>
> But even with this simplified approximation of the characteristic equation you can appreciate the value of having a characteristic curve in front of you to look at. If I had the option between having to use the equation or a diode characteristic curve to calculate the current through a diode, I know which I'd choose!

## Load lines

It's important to remember that although a diode characteristic curve is non-linear and non-ohmic, so that it doesn't abide by Ohm's law throughout its entire length, it does follow Ohm's law at any particular point on the curve. So, say, if the voltage across the diode whose characteristic curve is shown in Figure 7.1 is 0V8, so the current through it is about 80 mA (check it yourself) then the diode resistance is:

$$R = \frac{V}{I} = \frac{0.8}{80 \times 10^{-3}} = 10\Omega$$

as defined by Ohm's law. Any change in voltage and current, however, results in a different resistance.

Generally, diodes don't exist in a circuit merely by themselves. Other components e.g. resistors, capacitors, and other components in the semiconductor family, are combined with them. It is when designing such circuits and calculating the operating voltages and currents in the circuits that the use of diode characteristic curves really come in handy. Figure 7.4 shows as an example a simple circuit consisting of a diode, a resistor and a battery. By looking at the circuit we can see that a current will flow. But what is this current? If we knew the voltage across the resistor we could calculate (from Ohm's law) the current through it, which is of course the circuit current. Similarly, if we knew the voltage across the diode we could determine (from the characteristic curve) the circuit current. Unfortunately we know neither voltage!

We do know, however, that the voltages across both the components must add up to the battery voltage. In other words:

$$V_B \;=\; V_D + V_R$$

(it's a straightforward voltage divider). This means that we

**Figure 7.4    A simple diode circuit – but what is the current flow?**

can calculate each voltage as being a function of the battery voltage, given by:

$$V_D = V_B - V_R$$

and

$$V_R = V_B - V_D$$

Now, we know that the voltage across the diode can only vary between about 0V and 0V8 (given by the characteristic curve), but there's nothing to stop us hypothesising about larger voltages than this, and drawing up a table of voltages which would thus occur across the resistor. Table 7.1 is such a table, but it takes the process one stage further by calculating the hypothetical current through the resistor at these hypothetical voltages.

From Table 7.1 we can now plot a second curve on the diode characteristic curve, of diode voltage against resistor current. Figure 7.6 shows the completed characteristic curve (labelled Load line R=60R). The curve is actually a straight line — fairly obvious, if you think about it, because all we've done is plot a voltage and a current for a resistor, and resistors are ohmic and linear. Because in such a circuit as that of Figure 7.4 the resistor is known as a load i.e. it absorbs electrical power, the line on the characteristic curve representing diode voltage and resistor current is called the load line.

Where the load line and the characteristic curve cross is the operating point. As its name implies, this is the point representing the current through and voltage across the components in the circuit. In this example the diode voltage is thus 1 V and the diode current is 33 mA at the operating point.

| Diode voltage (VD) | Resistor voltage $(V_B - V_D)$ | Resistor current (I) |
|:---:|:---:|:---:|
| 3 | 0 | 0 |
| 2.5 | 0.5 | 8.3 mA |
| 2.0 | 1.0 | 16.7 mA |
| 1.5 | 1.5 | 25 mA |
| 1.0 | 2.0 | 33.3 mA |
| 0.5 | 2.5 | 41.7 mA |
| 0 | 3.0 | 50 mA |

**Table 7.1   Diode characteristics**

If you think carefully about what we've just seen, you should
see that the load line is a sort of resistor characteristic curve.
It doesn't look quite like those of Figure 7.3, however, because
the load line does not correspond to resistor voltage and
resistor current, but diode voltage and resistor current — so
it's a sort of inverse resistor characteristic curve — shown
in Figure 7.5.

**Figure 7.5   Load line from Table 1 results**

153

# Starting electronics

The slope of the load line (and thus the exact position of the circuit operating point) depends on the value of the resistor. Let's change the value of the resistor in Figure 7.4 to say 200 Ω. What is the new operating point? Draw the new load line corresponding to a resistor of value 200 Ω on Figure 7.5 to find out. You don't need to draw up a new table as in Table 7.1 — we know it's a straight line so we can draw it if we have only two points on the line. These two points can be when the diode voltage is 0 V (thus the resistor voltage:

$$V_R = V_B - V_D$$

equals the battery voltage), and when the diode voltage equals the battery voltage and so the resistor current is zero. Figure 7.6 shows how your results should appear. The new operating point corresponds to a diode voltage of 0V8 and current of about 11 mA. We'll be looking at other examples of the uses of load lines when we look at other semiconductor devices in later chapters.

**Figure 7.6    This shows the new load line for a resistor of 200 Ω**

154

# Diodes II

## Diode circuits

We're now going to look at some ways that diodes may be used in circuits for practical purposes. We already know that diodes may be used in circuits for practical purposes. We already know that diodes allow current flow in only one direction (ignoring saturation reverse current and zener current for the time being) and this is one of their main uses — to rectify alternating current (a.c.) voltages into direct current (d.c.) voltages. The most typical source of a.c. voltage we can think of is the 230 V mains supply to every home. Most electronic circuits require d.c. power so we can understand that rectification is one of the most important uses of diodes. The part of any electronic equipment — TVs, radios, hi-fis, computers — which rectifies a.c. mains voltages into low d.c. voltages is known as the power supply (sometimes abbreviated to PSU — for power supply unit).

Generally, we wouldn't want to tamper with voltages as high as mains, so we would use a transformer to reduce the 230 V a.c. mains supply voltage to about 12 V a.c. We'll be looking at transformers in detail in a later part, but suffice it to know now, that a transformer consists essentially of two coils of wire which are not in electrical contact. The circuit symbol of a transformer (Figure 7.7) shows this.

The simplest way of rectifying the a.c. output of a transformer is shown in Figure 7.8. Here a diode simply allows current to flow in one direction (to the load resistor, $R_L$ but not in the other direction (from the load resistor). The a.c. voltage from the transformer and the resultant voltage across the load resistor are shown in Figure 7.9. The resistor voltage, although in only one direction, is hardly the fixed voltage we would like,

155

Figure 7.7    The circuit symbol for a transformer

Figure 7.8    A simple output rectifier circuit using a diode

but nevertheless is technically a d.c. voltage. You'll see that of each wave or cycle of a.c. voltage from the transformer, only the positive half gets through the diode to the resistor. For this reason, the type of rectification shown by the circuit Figure 7.8 is known as half-wave rectification.

It would obviously give a much steadier d.c. voltage if both half waves of the a.c. voltage could pass to the load. We can do this in two ways. First by using a modified transformer, with a centre-tap to the output or secondary coil and two diodes as in Figure 7.10. The centre-tap of the transformer

156

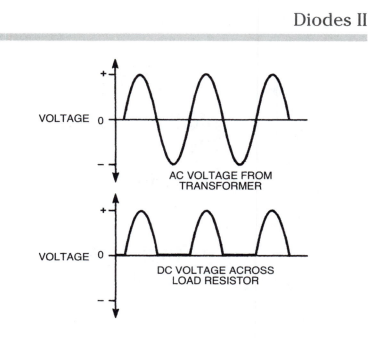

VOLTAGE 0

AC VOLTAGE FROM
TRANSFORMER

VOLTAGE 0

DC VOLTAGE ACROSS
LOAD RESISTOR

**Figure 7.9    Waveforms showing the output from the transformers, and the
rectified d.c. voltage across the load resistor**

gives a reference voltage to the load, about which one of the
two ends of the coil must always have a positive voltage (i.e.
if one end is positive the other is negative, if one end is negative the other must be positive) so each half-wave of the a.c.
voltage is rectified and passed to the load. The resultant d.c.
voltage across the load is shown.

Second, an ordinary transformer may be used with four
diodes as shown in Figure 7.11. The group of four diodes is
often called a bridge rectifier and may consist of four discrete
diodes or can be a single device which contains four diodes
in its body.

Figure 7.10 A more sophisticated rectifier arrangement using two diodes and a transformer centre tap

Both of these methods give a load voltage where each half wave of the a.c. voltage is present and so they are known as full-wave rectification.

Figure 7.11 A familiar rectifier arrangement: four diodes as a bridge rectifier

## Filter tips

Although we've managed to obtain a full-wave rectified d.c. load voltage we still have the problem that this voltage is not too steady (ideally we would like a fixed d.c. voltage which doesn't vary at all). We can reduce the up-and-down variability of the waves by adding a capacitor to the circuit output. If you remember, a capacitor stores charges — so we can use it to average out the variation in level of the full-wave rectified d.c. voltage. Figure 7.12 shows the idea and a possible resultant waveform. This process is referred to as smoothing and a capacitor used to this effect is a smoothing capacitor. Sometimes the process is also called filtering.

Figure 7.12   Levelling the rectifier d.c. with the help of a capacitor: the process is known as smoothing

159

You should remember that the rate at which a capacitor discharges is dependent on the value of the capacitor. So, to make sure the stored voltage doesn't fall too far in the time between the peaks of the half cycles, the capacitor should be large enough (typically of the value of thousands of microfarads) to store enough charge to prevent this happening. Nevertheless a variation in voltage will always occur, and the extent of this variation is known as the ripple voltage, shown in Figure 7.13. Ripple voltages of the order of a volt or so are common, superimposed on the required d.c. voltage, for this type of circuit.

Figure 7.13    The extent to which the d.c. is not exactly linear is known as the ripple voltage

## Stability built in

The power supplies we've seen so far are simple but they do have the drawback that output voltage is never exact — ripple voltage and to a large extent, load current requirements mean that a variation in voltage will always occur. In many practical applications such supplies are adequate, but some applications require a much more stable power supply voltage.

We've already seen a device capable of stabilising or regulating power supplies — the zener diode. Figure 7.14 shows the simple zener circuit we first saw last chapter. You'll remember that the zener diode is reverse biased and maintains a more or less constant voltage across it, even as the input voltage $V_{in}$ changes. If such a zener circuit is used at the output of a smoothed power supply (say, that of Figure 7.12) then the resultant stabilised power supply will have an output voltage which is much more stable, with a much reduced ripple voltage.

**Figure 7.14    A simple zener circuit which can be used with a smoothing circuit to give a greatly reduced ripple voltage**

Zener stabilising circuits are suitable when currents of no more than about 50 mA or so are required from the power supply. Above this it's more usual to build power supplies using stabilising integrated circuits (ICs), specially made for the purpose. Such ICs, commonly called voltage regulators, have diodes and other semiconductors within their bodies which provide the stabilising stage of the power supply. Voltage regulators give an accurate and constant output voltage with extremely small ripple voltages, even with large variations in load current and input voltage. ICs are produced which can provide load current up to about 5 A.

# Starting electronics

In summary, the power supply principle is summarised in Figure 7.15 in block diagram form. From a 230 V a.c. input voltage, a stabilised d.c. output voltage is produced. This is efficiently done only with the use of diodes in the rectification and stabilisation stages.

**Figure 7.15    A block diagram of a power supply using the circuit stages we have described**

## Practically there

Figure 7.16 shows a circuit we've already seen and built before: a 555-based astable multi-vibrator. We're going to use it to demonstrate the actions of some of the principles we've seen so far. Although the output of the astable multi-vibrator is a squarewave, we're going to imagine that it is a sinewave such

**Figure 7.16    An astable multi-vibrator circuit. This can be used to demonstrate some of the principles under discussion**

162

as that from a 230 V a.c. mains powered transformer. Figure 7.17(a) shows the squarewave output, and if you stretch your imagination a bit (careful, now!), rounding off the tops and bottoms of the waveform, you can approximate it to an alternating sinewave.

**Figure 7.17(a)  The squarewave output from the circuit in Figure 7.16 experimentally modified by a 100 kΩ resistor (b) and a 10 kΩ resistor (c)**

Build the circuit, as shown in the breadboard layout of Figure 7.18, and measure the output voltage. It should switch between 0 V and about 8 V.

Now let's add a smoothing capacitor $C_s$ and load resistor $R_L$ to the output of our imaginary rectified power supply, as in Figure 7.19. The output voltage of the smoothing stage is

**Figure 7.18    The breadboard layout of the circuit in Figure 7.16**

now fairly steady at about 6V5 indicating that smoothing has occurred. Although you can't see it, the voltage waveform would look something like that in Figure 7.17(b), with a ripple voltage of about 1 V. The breadboard layout for the circuit is shown in Figure 7.20.

We can show what happens when load current increases by putting a lower value load resistor in circuit — take out the 100 kΩ resistor and put a 10 kΩ resistor in its place. The ripple voltage now increases causing greater changes in the d.c. output level, and the whole waveform is shown as Figure 7.17(c).

**Figure 7.19   The same circuit with a smoothing capacitor added**

**Figure 7.20   The breadboard layout for the circuit in Figure 7.19**

That's all for this chapter. Next chapter we'll move onto another semiconductor — possibly the most important semiconductor ever — the transistor. Try the quiz over the page to see what you've learned here.

165

# Starting electronics

## Quiz

Answers at end of book

1 Resistor characteristic curves are:

   a nearly always linear
   b always exponential
   c nearly always exponential
   d rarely drawn
   e b and d
   f all of these

2 Diodes are:

   a ohmic
   b non-ohmic
   c linear
   d non-linear
   e a and c
   f b and d
   g none of these

3 The voltage across the diode whose characteristic curve is shown in Figure 7.1 is 0V7. What is the current through it?

   a 60 mA
   b about 700 mV
   c about 6 mA
   d it is impossible to say, because diodes are non-ohmic
   e none of these

4 Mains voltages are dangerous because:

   a they are a.c.
   b they are d.c.
   c they are high
   d they are not stabilised
   e none of these

5 A single diode cannot be used to full-wave rectify an a.c. voltage:

   True or false?

6 For full-wave rectification of an a.c. voltage:

   a a bridge rectifier is always needed
   b as few as two diodes can be used
   c an IC voltage regulator is essential
   d a and c
   e all of these
   f none of these

# 8 Transistors

In previous chapters we've looked at semiconductors, spending some time on diodes (the simplest semiconductor device) and taking brief glimpses of integrated circuits. It's the turn of the transistor now to come under the magnifying glass — perhaps the most important semiconductor of all.

You don't need many components this chapter, just a handful of resistors:

●     1 x 100 kΩ

●     1 x 220 kΩ

●     1 x 47 kΩ miniature horizontal preset

and, of course, one transistor.

The type of transistor you need is shown in Photo 8.1 and is identified as a 2N3053.

**Photo 8.1    A transistor next to a UK fifty pence piece**

You'll see that the transistor has three terminals called, incidentally, base, collector and emitter (often shortened to B, C and E). When you use transistors in electronic circuits it is essential that these three terminals go the right way round. The 2N3053 transistor terminals are identified by holding the transistor with its terminals pointing towards you from the body and comparing the transistor's underneath with the diagram in Figure 8.1. The terminal closest to the tab on the body is the emitter, then in a clockwise direction are the base and the collector.

Other transistor varieties may have different body types so it's important to check with reference books or manufacturer's data regarding the transistor terminals before use. All 2N3053 transistors, however, are in the same body type — known as a TO-5 body — and follow the diagram in Figure 8.1.

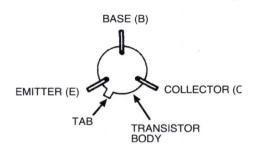

BASE (B)

EMITTER (E)

COLLECTOR (C

TAB

TRANSISTOR
BODY

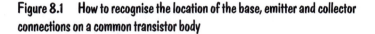

**Figure 8.1 How to recognise the location of the base, emitter and collector connections on a common transistor body**

For the previous two chapters, we've looked closely at diodes — the simplest of the group of components known as semiconductors. The many different types of diodes are all formed by combining doped layers of semiconductor material at a junction. The PN junction (as one layer is N-type semiconductor material and the other layer is P-type) forms the basis of all other semiconductor-based electronic components. The transistor — the component we're going to look at now — is, in fact, made of two PN junctions, back to back. Figure 8.2 shows how we may simply consider a transistor as being two back-to-back diodes, and we can verify this using our multimeter to check resistances between the three terminals of a transistor: the base, emitter and collector.

To do the experiment, put a transistor into your breadboard, then use the meter to test the resistance between transistor terminals. Using our brains here (I know it's hard) we can work out that if there are three terminals there must be six different ways the meter leads can connect to pairs of the terminals. Table 8.1 lists all six ways, but the results column is left blank for you to fill in as you perform the experiment.

169

**Figure 8.2    A symbol for a transistor considered as two diodes, back to back**

It's not necessary for you to measure the exact resistance obtained between terminals, it's sufficient just to find out if the resistance is high or low.

Take note — Take note — Take note — Take note

One final point before you start — the casing of the transistor body (sometimes called the can) is metal and therefore conducts. When you touch the meter test probes to the transistor terminals make sure, therefore, they don't touch the can. It's all too easy to do without realising you've got a short circuit which, of course, gives an incorrect result. The can of a TO-5 bodied transistor such as the 2N3053 is also electrically connected to the transistor's collector, so be careful!

| Black meter lead | Red meter lead | Result |
|---|---|---|
| Base | Emitter | |
| Emitter | Base | |
| Base | Collector | |
| Collector | Base | |
| Emitter | Collector | |
| Collector | Emitter | |

**Table 8.1 Six different ways to connect the meter leads**

Your results should show that low resistances occur in only two cases, indicating forward current flow between base and emitter, and base and collector. This corresponds as we should expect to the diagram of Figure 8.2.

## Very close

Unfortunately things aren't quite as simple as that in electronic circuits, where individual resistances are rarely considered alone. The real life transistor deals with currents in more than one direction and this confuses the issue. However, all we need know here (thankfully) is that the two PN junctions are very close together. So close that, in fact, they affect one another. It's almost as if they're Siamese twins — and whatever happens to one affects the other.

Figure 8.3 illustrates how a transistor can be built up, from two PN junctions situated very close together. It's really only a thin layer of P-type semiconductor material (only a few hundred or so atoms thick) between two thicker layers of N-type

**Figure 8.3   A narrow P-area gives two PN junctions very close together**

semiconductor material. Now let's connect this transistor arrangement between a voltage supply, so that collector is positive and emitter is zero, as in Figure 8.4.

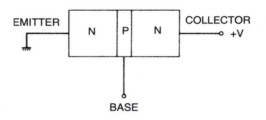

**Figure 8.4   Connected up, the emitter is at 0V and the collector is positive**

From what we know so far, nothing can happen and no current can flow from collector to emitter because between these two terminals two back-to-back PN junctions lie. One of these junctions is reverse biased and so, like a reverse biased diode, cannot conduct. So, what's the point of it all?

Well, as mentioned earlier, what happens in one of the two PN junctions of the transistor affects the other. Let's say for example that we start the lower PN junction (between base and emitter) conducting by raising the base voltage so that the base-to-emitter voltage is above the transition voltage of the junction (say, 0V6 if the transistor is a silicon variety). Figure 8.5 shows this situation. Now, the lower junction is flooded with charge carriers and because both junctions are very close together, these charge carriers allow current flow from collector to emitter also, as shown in Figure 8.6.

**Figure 8.5** If the voltage at the base is raised, current flows from base to emitter

**Figure 8.6** Charge carriers accumulating around the lower junction allow current to flow from the collector to the emitter

173

So, to summarise, a current will flow from collector to emitter of the transistor when the lower junction is forward biased by a small base-to-emitter voltage. When the base-to-emitter voltage is removed the collector-to-emitter current will stop.

We can build a circuit to see if this is true, as shown in Figure 8.7. Note the transistor circuit symbol. Figure 8.8 shows the breadboard layout. From the circuit you'll see that we're measuring the transistor's collector-to-emitter current (commonly shortened to just collector current) when the base-to-emitter junction is first connected to zero volts and to all intents and purposes is reverse biased, then when the base is connected to positive and the base-emitter junction is forward biased.

Now do the experiment and see what happens. You should find that when the base is connected to zero via the 100 kΩ resistor nothing is measured by the multimeter. But when the base resistor is connected to positive, the multi-meter shows a collector current flow. In our experiment

Figure 8.7    An experimental circuit to test what we have described so far

**Figure 8.8    A breadboard layout for the circuit in Figure 8.7**

a collector current of about 12 mA was measured — yours may be a little different. Finally, when the base resistor is returned to zero volts (or simply when it's disconnected!) the collector current again does not flow.

So what? What use is this? Not a lot as it stands, but it becomes very important when we calculate the currents involved. We already know the collector current (about 12 mA in our case) but what about the base-to-emitter current (shortened to base current)? The best way to find this is not by measurement (the meter itself would affect the transistor's operation!) but by calculation. We know the transistor's base-to-emitter voltage (shortened to base voltage) and we know the supply voltage.

175

# Starting electronics

From these we can calculate the voltage across the resistor, and from Ohm's law we can therefore calculate the current through the resistor. And the current through the resistor must be the base current.

The resistor voltage is:

$$9 - 0.7 = 8.3\text{\textbackslash}$$

So, from Ohm's law the current is:

$$I = \frac{V}{R} = \frac{8.3}{100,000} = 8.3\mu A$$

Not a lot!

Now we can begin to see the importance of the transistor. A tiny base current can turn on or off a quite large collector current. This is illustrated in Figure 8.9 and is of vital importance — so remember it!

**Figure 8.9** Important: a very small base current can control a large collector current

176

In effect, the transistor is a current amplifier. No matter how small the base current is, the collector current will be much larger. The collector current is, in fact, directly proportional to the base current. Double the base current and you double the collector current. Halve the base current and the collector current is likewise halved. It's this fact that the transistor may be used as a controlling element (the base current controlling the collector current) which makes it the most important component in the electronics world.

The ratio:

$$\frac{\text{collector current}}{\text{base current}}$$

gives a constant of proportionality for the transistor, which can have many names depending on which way you butter your bread. Officially it's called the forward current transfer ratio, common emitter, but as that's quite a mouthful it's often just called the transistor's current gain (seems sensible!). You can sometimes shorten this even further if you wish, to the symbols: $h_{fe}$, or ß. In manufacturer's data sheets for transistors the current gain is normally just given the symbol $h_{fe}$ (which if you're interested stands for Hybrid parameter, Forward, common Emitter. Are you any wiser?). However, we'll just stick to the term current gain, generally.

We can work out the current gain of a transistor by measuring the collector current, and calculating the base current as we did earlier, and dividing one by the other. For example the current gain of the transistor we used is:

$$\frac{12 \times 10^{-3}}{8.3 \times 10^{-6}} = 145$$

Yours may be a bit different. Manufacturers will quote typical values of current gain in their data sheets — individual
transistors' current gains will be somewhere around this
value, and may not be exact at all. It really doesn't matter,
too much. The transistor we use here, the 2N3053, is a fairly
common general-purpose transistor. High power transistors
may have current gains more in the region of about 10, while
some modern transistors for use in high frequency circuits
such as radio may have current gains around 1000 or so.

## NPN

The 2N3053 transistor is known as an NPN transistor because
of the fact that a thin layer of P-type semiconductor material
is sandwiched between two layers of N-type semiconductor material. The construction and circuit symbol of an
NPN transistor are shown in Figure 8.10(a). You may have
worked out that there is another way to sandwich one type of
semiconductor material between two others — and if so you
would also have worked out that such a transistor would be
called a PNP transistor. Figure 8.10(b) shows a PNP transistor construction and its circuit symbol. The only difference
in the circuit symbols of both types is that the arrow on an
NPN transistor's emitter points out and the arrow on a PNP
transistor's emitter points in.

The emitter arrow of either symbol indicates direction of
base current and collector current flow. So from the circuit
symbols we can work out that base current in the NPN tran-

(a) NPN                                    (b) PNP

**Figure 8.10    The internal construction and circuit symbols for NPN and PNP transistors**

sistor flows from base-to-emitter, while in the PNP transistor it flows from emitter-to-base. Likewise collector current flow in the NPN transistor is from collector-to-emitter and from emitter-to-collector in the PNP transistor.

Knowing this and comparing the PNP construction to that of the NPN transistor we can further work out that a tiny emitter-to-base current (still called the base current, incidentally) will cause a much larger emitter-to-collector current (still called the collector current). This is illustrated in Figure 8.11. The ratio of collector current to base current of a PNP transistor is still the current gain. In fact, apart from the different directions of currents, a PNP transistor functions identically to an NPN transistor. As we've started our look at transistors with the use of an NPN transistor, however, we'll finish it the same way.

179

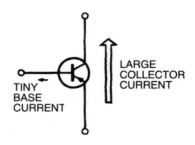

TINY BASE CURRENT

LARGE COLLECTOR CURRENT

**Figure 8.11** An NPN has the same effect as a PNP transistor, but in the opposite direction

## Using transistors

We've seen how transistors work but we don't yet know how they can be used. After all, there are millions and millions of transistors around in the world today — you'd be forgiven for thinking that there must be hundreds, if not thousands of ways that a transistor may be used.

Well, you'd be forgiven, but you'd still be wrong!

In fact, to prove the point we're now going to go through every possible way a transistor can be made to operate, and as this chapter isn't five miles thick you've probably realised that there aren't many ways a transistor may be used at all. Incredibly, at the last count, the number of ways a transistor may be used is — dah, da, dah, dah, dah, da, dah, dah — two!

> **Hint:**
>
> Yes, that's right, only two basic uses of a transistor exist, and every transistorised circuit, every piece of electronic equipment, every television, every radio, every computer, every digital watch and so on, contains transistors in one form or another which do only one of two things.

We've already seen the first of these two uses — an electronic switch, where a tiny base current turns on a comparatively large collector current. This may appear insignificant in itself, but if you consider that the collector current of one transistor may be used as the base current of a following transistor or transistors, then you should be able to imagine an enormous number of transistors inside one appliance, all switching and hence controlling the appliance's operation.

Incredible? Science Fiction? Well, let me tell you that the appliance described here in extremely simple terms already exists, in millions. We call the appliance a computer. And every computer contains thousands if not millions of transistor electronic switches.

We can see the second use of a transistor in the circuit of Figure 8.12. From this you should be able to see that we're going to use a variable resistor to provide a variable base current for the transistor in the circuit. The breadboard layout of the circuit is in Figure 8.13. Before you connect the battery, make sure the preset variable resistor is turned fully anticlockwise.

# Starting electronics

**Figure 8.12    A circuit to test a transistor with a varible base current**

**Figure 8.13    A breadboard layout for the circuit in Figure 8.12. The preset should be turned slowly with a fine screwdriver**

Now, connect the battery and slowly (with a small screwdriver) turn the preset clockwise. Gradually, as you turn the preset, the LED should light up: dim at first, then brighter, then fully bright. What you've built is a very simple lamp dimmer.

So, the other use of a transistor is as a variable control element. By controlling the transistor's tiny base current we can control the much larger collector current, which can be the driving current of, say, a LED, an ordinary lamp, a motor, a loudspeaker — in fact just about anything which is variable.

182

These two operational modes of transistors have been given names. The first, as it switches between two states, one where the collector current is on or high, the other where it is off or low, is called digital. Any circuit which uses transistors operating in digital mode is therefore called a digital circuit.

The other mode, where transistors control, is known as the analogue mode, because the collector current of the transistor is simply an analogue of the base current. Any circuit which uses transistors operating in the analogue mode is known as an analogue circuit.

> ## Take note — Take note — Take note — Take note
>
> Sometimes analogue circuits are mistakenly called linear circuits, However, this is wrong, because although it might appear that a linear law is followed, this is not so. If transistors were properly linear they would follow Ohm's law, but (like diodes) — remember, they do not follow Ohm's law and so are non-linear devices. In an analogue circuit, however, transistors are operated over a part of their characteristic curve (remember what a characteristic curve is from Chapters 6 & 7?) which approximates a straight line — hence the mistaken name linear. Of course, you won't ever make the mistake of calling an analogue circuit linear, will you?

Yes, like diodes, transistors have characteristic curves too, but as they have three terminals they have correspondingly more curves.

# Starting electronics

## Quiz

Answers at end of book

1  In the circuit symbol for an NPN transistor:

   a  the arrow points out of the base
   b  the arrow points into the base
   c  the arrow points out of the collector
   d  the arrow points into the collector
   e  b and c
   f  all of these
   g  none of these.

2  In an NPN transistor:

   a  a small base current causes a large collector current to flow
   b  a small collector current causes a large base current to flow
   c  a small emitter current causes nothing to flow
   d  nothing happens
   e  none of these
   f  all of these.

3  If an NPN transistor has a current gain of 100, and a base current of 1 mA, its collector current will be:

   a  1 mA
   b  1 $\mu$A
   c  10 $\mu$A
   d  100$\mu$A.
   e  bigger

   f  none of these.

4  The current gain of a transistor:

   a  is equal to the collector current divided by the emitter current

   b  is equal to the collector current divided by the base current

   c  is equal to the base current divided by the collector current

   d  a and c
   e  all of these
   f  none of these.

5  The current gain of a transistor has units of:

   a  amps
   b  milliamps
   c  volts
   d  this is a trick question, it has no units
   e  none of these.

# 9 Analogue integrated circuits

Well, the last chapter was pretty well jam-packed with information about transistors. If you remember, we saw that transistors are very important electronic components. They may be used in one of only two ways: in digital circuits, or in analogue circuits — although many, many types of digital and analogue circuits exist.

In terms of importance, transistors are the tops. They're far more important than, say, resistors or capacitors, although in most circuits they rely on the other components to help them perform the desired functions. They're the first electronic components we've come across which are active: they actually control the flow of electrons through themselves, to perform their function — resistors and capacitors haven't got this facility, they are passive and current merely flows or doesn't flow through them.

Transistors, being active, can control current so that they can be turned into amplifiers or switches depending on the circuit. There's nothing magical about this, mind you, we're not gaining something for nothing! In order that, say, a transistor can amplify a small signal into a large one, energy has to be added in the form of electricity from a power supply. The transistor merely controls the energy available from this power supply, creating the amplification effect.

They're important for another reason, however. They are small! They can be made by mass-production techniques, almost as small as you can imagine, and certainly many times smaller than you could see with your bare eyes. This in itself is no big deal — imagine trying to solder a transistor into a circuit which was so small you couldn't even see it! Transistors of the types we've seen so far are purposefully made as large as they are, just so we can handle them.

## Integrated circuits

But the big advantage of transistors, especially tiny transistors, is that many of them may be integrated into a single circuit, which is itself still very small. To give you an idea of what we're talking about, modern integrated circuits (well, what else would you call them?) have been made with over a quarter of a million transistors, all fitting onto one small silicon chip only about forty square millimetres or so! Now this sort of integration represents the ultimate in human achievement, remember, but processes are being improved every year and integrated

# Analogue integrated circuits

circuits with only hundreds of thousands of integrated transistors are now commonplace. Because of this it stands to reason that we have to come to grips with integrated circuits, they're here to stay and there's no point in being shy. So the final chapters of this epic saga are devoted to integrated circuits.

The integrated circuit which we are going to play with this chapter is an analogue integrated circuit (let's cut all these long words out and call them by their abbreviation — IC). Just as there are analogue and digital circuits which transistors are used in, so there are analogue and digital ICs. This IC just happens to be analogue, but there are digital ones too — the 555 (which we've already seen) is actually an example. In IC terms the IC in this chapter is a comparatively simple IC, having only around twenty transistors in all; nevertheless it is an extremely versatile IC and is probably the most commonly used IC of all time.

Generally it's called the 741, referring to the numbers which are printed on the top of the IC body — see Photo 9.1. If you look at your IC you'll see that it also has some letters associated with the number 741. These letters e.g., LM, SN, $\mu$, MC, refer to the manufacturer of the device and bear no relevance to the internal circuit, which is the same in all cases. An SN741 is identical in performance to a $\mu$741. Other letters printed on the top of the IC refer to other things such as batch number, date of manufacture etc. Whatever is printed on the top of the IC, as long as it is a 741 there's normally no problem and it'll work in all the circuits we're going to look at now.

# Starting electronics

**Photo 9.1    Our subject: a 741 op-amp**

You'll notice that the 741 is identical in physical appearance to the other IC we've already looked at, the 555. Like the 555, the 741 is housed in an 8-pin DIL (dual-in-line) body. But although the shape's the same, the internal circuit isn't. Other versions of the 741 exist, in different body styles, but the 8-pin DIL version is by far the most popular.

The 741's internal circuit is shown in Figure 9.1. Its circuit symbol is shown in Figure 9.2. From this you can see that the circuit has two inputs, marked + and –. Technically speaking these are called the non-inverting and inverting inputs. The circuit has one output and two inputs for power supply connection. There are also two other connectors to the circuit, known as offset null connections; controlling the voltages of these inputs, controls the voltage level of the output — you'll find out how these connections work soon. They are not used in every circuit which the 741 is used in.

Figure 9.1    The internal circuit of the 741: a mere 22 transistors

Figure 9.2    The circuit symbol for a 741

The internal layout of the 8-pin DIL version of the 741 is shown in Figure 9.3. Now, each input and output of the op-amp is associated with a particular pin of the DIL body. This means that we can redraw the circuit symbol of the 741, as in Figure 9.4, with the corresponding pin numbers at each connection.

189

OFFSET NULL 1 — 8 NO CONNECTION
INVERTING INPUT 2 — 7 POSITIVE SUPPLY
NON-INVERTING INPUT 3 — 6 OUTPUT
NEGATIVE SUPPLY 4 — 5 OFFSET NULL

**Figure 9.3 The internal layout of an 8-pin DIL-packed 741**

Note that these pin numbers are only correct when we deal with the 8-pin DIL version, however, and any other versions would have different pin numbers. We'll show the 8-pin DIL pin numbers with any circuit diagram we show here, though, as we assume that you will use this version.

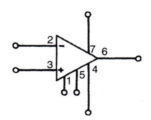

**Figure 9.4 The symbol in Figure 9.2, with the pin numbers included**

Apart from the 741, there are a small number of other components you'll also need for the circuits in this chapter. These are:

- 2 x 10 kΩ resistors

- 3 x 47 kΩ resistors

- 1 x 22 kΩ resistor

- 1 x 47 kΩ miniature horizontal preset

## Building on blocs

The first circuit is shown in Figure 9.5, and from this you will see two 9 V batteries are used as a power supply. This is because the 741 needs what's called a three-rail power supply i.e., one with a positive rail, a negative rail, and a ground rail 0 V. Now you may ask how two batteries can be used to provide this. Well the answer's simple, join them in series. This gives a +18 V supply, a 0 V supply, and at the mid-point between the batteries, a +9 V supply. As voltages are only relative anyway, we can now consider the +9V supply to be the middle ground rail of our three-rail power supply and refer to it as 0 V. The +18 V supply then becomes the positive rail (+9 V) and the 0 V supply becomes the negative rail (–9 V).

### Hint:

The 741 is commonly referred to as an operational amplifier — shortened to op-amp. The term comes from the earlier days in electronics when discrete components were built into circuits and used as complete general-purpose units. An op-amp, as its name suggests, is a general-purpose amplifier which is more-or-less ready-to-use. Years ago it might have been a large metal box full of valves and wiring. Later it would have been a circuit board on which the circuit was built with transistors. But nowadays nobody uses op-amps made with discrete components, because modern technology has produced IC op-amps such as the 741. Given the choice, would you use discrete components to build an op-amp of the complexity of that in Figure 9.1, or would you use a 741?

**Figure 9.5    Our first experimental circuit: a non-inverting amplifier with a three-rail power supply**

Besides the ease of use inherent with ICs, they are also cheaper than discrete counterparts — the 741 is priced at around 40 pence. Building the circuit using transistors would cost many times this.

As the 741 is an operational amplifier, it makes sense that the first circuit we look at is an amplifying circuit. But what sort of amplifier does it form? Well the easiest way to find out is to build the circuit, following the breadboard layout of Figure 9.6, and then use your multi-meter to measure the voltage at the circuit's input (pin 3) and compare it with the voltage at the circuit's output (pin 6). The voltages should be measured for a range of values, adjusted by the preset variable resistor RVl, and should be written into Table 9.1 as you measure them.

Figure 9.6    The breadboard layout for the circuit in Figure 9.5. Note that in this diagram the input voltage is being measured

| Input voltage | Output voltage |
|:---:|:---:|
| -2 | |
| -1 | |
| 0 | |
| 1 | |
| 2 | |

Table 9.1    To show the results of your measurements

Note that in Table 9.1 there are measurement points at negative as well as positive voltages: the best way of taking the readings is to first take the positive readings then turn the

# Starting electronics

meter round i.e., swap the meter leads over so that the black lead measures the voltage at pin 6 and the red lead connects to 0 V, to measure the negative readings.

The results which were taken in my circuit are given in Table 9.2, and your results should be something like this, although differences may occur. From these results you should see that the circuit amplifies whatever voltage is present at the input by about two, so that the output voltage is about twice that of the input. This is true whatever the input voltage. Don't worry if your results aren't that close to the ideal ones, it's only an experiment.

| Input voltage | Output voltage |
|---|---|
| -2 | -4.4 |
| -1 | -2.1 |
| 0 | 0 |
| 1 | 2.1 |
| 2 | 4.4 |

**Table 9.2    The results of my measurements**

This particular circuit is a fairly basic amplifier which simply amplifies the input voltage. For one reason or another, which you'll appreciate soon, it's called a non-inverting amplifier. Its gain is given by the expression:

$$G = \frac{R1}{R2} + 1$$

and if you insert the values of resistors R1 and R2 into this expression you will see why the circuit gives an experimental gain of about two.

As the circuit has a gain which is dependent on the values of resistors R1 and R2, it is a simple matter to adjust the circuit's gain, by changing one or both resistors. Try changing resistor R1 to say 22 k and see what the gain is (it should be about three).

---

**Take note — Take note — Take note — Take note**

Theoretically, of course, the circuit could have any gain we desire, just by inserting the right values of resistances. But in practice there are limits. If, say, the circuit's gain was set to be 20, and an input voltage of 1 V was applied, the circuit would attempt to produce an output voltage of 20 V. But that's not possible — the output voltage can't be greater than the power supply voltage (fairly obvious), which in this case is ±9 V. The answer, you may think, is to wind up the power supply voltage to + and –20 V, and, yes, you can do this to a certain extent, but again there are practical limitations. The 741, for example, allows a maximum power supply voltage of about ±15 V, above which it may be damaged.

---

## Turning things upside-down

Earlier, we saw how the op-amp has two inputs, which we called non-inverting and inverting inputs. Now if we used the non-inverting input of the op-amp to make a non-inverting amplifier (Figure 9.5) then it makes sense that the inverting input of the op-amp may be used to make an inverting amplifier, as shown in Figure 9.7.

**Figure 9.7    Our next experimental circuit: an inverting amplifier**

Build the circuit up on your breadboard, as shown in the breadboard layout of Figure 9.8. In a similar way to the last circuit; now take a number of voltage measurements of input and output voltages, to confirm that this really is an inverting amplifier. Remember that whatever polarity the input voltage is, the output voltage should be the opposite, so you'll have to change round the meter leads to suit.

### Hint:

One final point — the input to the op-amp circuit is classed as the connection between resistor R1 and the preset resistor, not the inverting input terminal at pin 2 — you'll get some peculiar results if you try to measure the input voltage at pin 2!

**Figure 9.8    The breadboard layout for the circuit in Figure 9.7**

Tabulate your results in Table 9.3. The results from my circuit are given in Table 9.4, for reference. You'll see that the circuit does invert the voltage, but also amplifies it by two at the same time.

| Input voltage | Output voltage |
|:---:|:---:|
| -2 | |
| -1 | |
| 0 | |
| 1 | |
| 2 | |

**Table 9.3    To show the results of your measurements**

197

Like the non-inverting amplifier, the inverting amplifier's gain is determined by resistance values, given this time by the expression:

$$G = \frac{R2}{R1}$$

so that the gain may once again be altered to suit a required application, just by changing a resistor.

| Input voltage | Output voltage |
|:---:|:---:|
| -2 | 4.2 |
| -1 | 2.1 |
| 0 | 0 |
| 1 | -2.1 |
| 2 | -4.2 |

**Table 9.4**  Shows the results of my measurements

## Follow me

These two op-amp formats; inverting and non-inverting amplifiers, are the basic building blocks from which most op-amp circuits are designed. They can be adapted merely by altering the values of the circuit resistances, or even by replacing the resistances with other components (such as capacitors) so that the circuits perform a vast number of possible functions.

As an example we'll look at a couple of circuits, considering how they are designed, compared with the two basic formats. First, we'll consider the circuit in Figure 9.9. At first sight we can see that it is a non-inverting circuit of some description — the input voltage (derived from the wiper of the preset resistor) is applied to the non-inverting input of the op-amp. But in comparison with our non-inverting amplifier of Figure 9.6, there are no resistors around the amplifier; instead the output voltage is simply looped back to the inverting input. So how does it work and what does it do?

**Figure 9.9   A voltage follower circuit: the output is equal to the input. This is basically a non-inverting circuit**

Let's first consider the non-inverting amplifier gain expression:

$$G = \frac{R1}{R2} + 1$$

and look to see how we can apply it here. Although no resistances are in this circuit that doesn't mean that we don't need to consider them! Resistor R1 in the circuit of Figure 9.5 can be seen to be between pin 6 of the op-amp (the output) and pin 2 (the inverting input). In the circuit of Figure 9.9, on the other hand, there is a direct connection between pin 6 and pin 2 — in other words we can imagine that resistor R1 of this circuit is zero.

Similarly, resistor R2 of the circuit in Figure 9.5 lies between pin 2 and 0 V. Between pin 2 and 0 V of the circuit in Figure 9.9 there is no resistor — in other words we can imagine the resistance to be infinitely high.

We can put these values of resistances into the expression for gain, which gives us:

$$G = \frac{0}{\infty} + 1$$

$$= 1$$

so the circuit has a gain of one i.e., whatever voltage is applied at the input is produced at the output. The circuit has a number of common names, one of which is voltage follower, which quite accurately describes its operation. Build the circuit as shown in the breadboard layout of Figure 9.10, to see for yourself that this is indeed what happens. All you need do to confirm the results is to measure the input and output voltages at a few different settings, making sure they are the same.

It may seem a bit odd that we have built a circuit, however simple, which appears to do the same job which an ordinary length of wire could perform — that is, the output voltage is the same as the input voltage. But the situation is not quite as simple as that. The circuit we have constructed, although having a gain of only one, has a very high input resistance. This means that very little current is drawn from the preceding circuit (in this case the preset resistor/voltage divider chain). On the other hand, the circuit has a very low output resistance, which means that it is capable of supplying a substantial current to any following circuits. This property gives rise to another of the circuit's common names — resistance converter.

**Figure 9.10    The breadboard layout of the circuit in Figure 9.9**

Hint:

In practice the voltage follower is often used when the output of a circuit which is only capable of producing a small, weak current is required to be amplified. The voltage follower is used in the input stage of the amplifier to buffer the current before amplification takes place, thus preventing loading of the preceding circuit. In fact the term buffer is just another name for the voltage follower.

# Starting electronics

## Offset nulling

Earlier on it was mentioned that there are two connections to the 741 which are called offset null connections. In most op-amp applications these would not be used, but in certain instances, where the level of the output voltage is of critical importance they are.

When the op-amp has no input voltage applied to it — or, more correctly speaking, the input voltage is 0 V — the output voltage should be the same i.e., 0 V. Under ideal conditions of manufacture this would be so, but as the 741 is a mass-produced device there are inevitable differences in circuit operation and so the output is rarely exactly 0 V. Temperature differences can also create changes in this output voltage level. The difference in the output from 0 V is known as the offset voltage and is usually in the order of just a few millivolts.

> **Hint:**
>
> In the majority of applications this level of offset voltage is no problem and so no action is taken to eliminate it. But the offset null terminals of the 741 may be used to control the level of offset voltage to reduce it to zero.

Figure 9.11 shows a circuit which you can build to see the effects of the process of offset nulling, where a potentiometer has been connected between the two offset null terminals

**Figure 9.11**   An experimental circuit to demonstrate the process of offset nulling

with its wiper connected to the negative supply. Adjusting the wiper controls the offset voltage. You'll see that the circuit is basically a buffer amplifier, like that of Figure 9.9, in which the input voltage to the non-inverting terminal is 0 V.

Figure 9.12 shows the breadboard layout of this offset nulling circuit. The multi-meter should be set to its lowest voltage setting e.g., 0.1 V, and even at this setting you won't notice much difference in the output voltage. In fact, you probably won't be able to detect any more than a change of about ±10 millivolts as you adjust the preset resistor.

## Get back

So far, we've used the 741 op-amp in a variety of circuits, all of which form the basis for us to build many, many more circuits. And that's fine if you only want to be given circuits

Figure 9.12  The breadboard layout of the circuit in Figure 9.11 on the previous page

which you can build yourself. But the more inquisitive reader may have been wondering how the device can do all of this: after all, it's only through knowledge that greater things can be achieved — to design your own circuits using op-amps, you need to know what makes them tick.

An op-amp can be considered to be a general-purpose amplifier with an extremely high gain. By itself (that is, with no components connected to it) it will have a gain of many thousands. Typical figures for the 741, for example, give a gain of 200,000 at zero frequency, although this varies considerably from device to device. Now, nobody's suggesting that this sort of gain is particularly useful in itself — can you think of an application in which a gain like this is required? — and besides, as each device has a different gain it would be well nigh impossible to build two circuits with identical properties, let alone the mass-produced thousands of radios, TVs, record players and so on which use amplifiers. So we need some way of taming this high gain, at the same time as defining its value precisely, so that useful and accurate amplifiers may be designed.

The process used in this taming of op-amps is known as feedback i.e., part of the output signal from the op-amp is fed back to the input. Look closely again at the circuits we have built so far this chapter and you'll see that in all cases there is some connection or other from the output back to the input. These connections form the necessary feedback paths which reduce the amplifier's gain to a determined, precise level.

# Starting electronics

## Quiz

Answers at end of book

1  A typical operational amplifier such as a 741 has:

    a  two inputs and one output
    b  two outputs and one input
    c  three inputs
    d  one input and one output
    e  c and d
    f  none of these

2  A voltage follower is:

    a  an op-amp circuit with a gain of 100 but which is non-inverting
    b  a unity gain inverting op-amp amplifier
    c  an inverting amplifier, built with an op-amp
    d  b and c
    e  all of these
    f  none of these

3  An integrated circuit has:

    a  many transistors
    b  an internal chip of silicon
    c  pins which allow connection to and from it
    d  all of these

    e  none of these

4  In the non-inverting amplifier of Figure 9.5, the gain of the circuit when resistor R1 is 100 k, is:

    a  1.1
    b  0.9
    c  12
    d  21
    e  none of these

5  The gain of an op-amp inverting amplifier can be 0.1:

    True or false

6  The gain of an op-amp non inverting amplifier can be 0.1:

    True or false

# 10 Digital integrated circuits I

If you take a close look at an electronic process — any electronic process — you will always find some part of it which is automatic. By this, I mean that some mechanism is established which controls the process to some extent without human involvement.

Electronic control can be based on either of the two electronics principles we've already mentioned: analogue circuits, or digital circuits. In the last chapter we saw how analogue circuits can be built into integrated circuits. This chapter concentrates on the use of integrated circuits which contain digital circuits — which we call digital integrated circuits.

# Starting electronics

As we've already seen, integrated circuits (whether analogue or digital) are comprised of transistors — lots of 'em. So it's to the transistor that we first turn to in this chapter's look at digital integrated circuits.

When we first considered transistors, we saw that they can operate in only one of two modes: analogue (sometimes, mistakenly called linear) where the transistor operates over a restricted portion of its characteristic curve, and digital (where the transistor merely acts as a electronic switch — which can be either on or off).

It should be no surprise to learn that analogue ICs contain transistors operating in analogue mode, while digital ICs contain transistors operating in switching mode. Just occasionally, ICs contain both analogue and digital circuits, but these are usually labelled digital anyway.

Figure 10.1 shows a transistor switch which shows the general operating principles. It's a single device, with an input at point A, and a resultant output at point B. How does it work? Well, we need to mull over a few terms first, before we can show this.

**Figure 10.1    A simple transistor used as a switch, to form a simple digital circuit**

## Logically speaking

In the electronic sense, digital signifies that a device has fixed states. Its input is either on or off, its output is also either on or off. As such there are only really two states, which incidentally gives rise to another term that is often used to describe such electronic circuits — binary digital. Also, it's common to know the states by more simple names, such as on and off, or sometimes logic 0 and logic 1.

---

**Hint:**

States in an electronic digital circuit are generally indicated by different voltages, with a particular voltage implying one state, and another voltage implying the other state. The actual voltages are irrelevant, as long as it's known which voltage implies which state, although by convention typical voltages might be 0 V when the state is off or logic 0, and 5 V when the state is on or logic 1. As long as there is a defined voltage for a defined state, it doesn't matter. Further, and as a general rule, just the numbers 0 and 1 can be used to distinguish between the two logic states — whatever the voltages used.

---

In terms of electronic logic, the circuit of Figure 10.1 is quite simple in operation, and we can work out what it does by considering the transistor's operation. We know from the experiment on page 174 that the presence of a small base current initiates a large collector current through the transistor. We can use this knowledge to calculate what happens in the circuit of Figure 10.1.

# Starting electronics

If the input at point A is logic 0 (we shall assume that logic 0 is represented by the voltage 0 V, and that logic 1 is represented by the voltage 10 V) then no base current flows, so no collector current flows — the transistor is switched off. This lack of collector current makes the transistor the equivalent of an extremely high resistance. An equivalent circuit representing this condition is shown in Figure 10.2, where the transistor resistance Rtr is 20 M (a typical value for a transistor in the off mode).

**Figure 10.2    Equivalent diagram of a single transistor switch, with the transistor base (point A in Figure 10.1) connected to logic 0**

Resistors Rtr and R1 effectively form a voltage divider, the output voltage of which follows the formula:

$$V_B = \frac{R_{tr}}{R_{tr} + R1} \times V_A$$
$$= \frac{20M}{20M + 1k} \times 10$$

which is close enough to being 10 V (that is logic 1) to make no difference.

The effect is that output B is connected to logic 1 — in other words, the output at point B is the opposite of that at point A.

If we now consider the transistor with a base current (that is, we connect point A to logic 1, at 10 V), the transistor turns on and passes a large collector current, with the effect that it become a low resistance. The equivalent circuit is shown in Figure 10.3.

**Figure 10.3    Equivalent diagram of a single transistor switch, with point A connected to logic 1**

The output voltage at point B is now:

$$V_B = \frac{10}{1k + 10} \times 10$$

which is as close to 0 V (logic 0) as we'll ever get.

So in this circuit the transistor output state is always the opposite of the input state. Put another way, whatever the input state, the output is always the inverse.

We call this simple digital circuit — you guessed it — an inverter.

# Starting electronics

## Every picture tells a story

Interestingly, digital circuits of all types (and we'll cover a few over the next few pages) are made up using just this simple and basic configuration. Even whole computers use this basic inverter as the heart of all their complex circuits.

To make it easy and convenient to design complex digital circuits using such transistor switches, we use symbols instead of having to draw the complete transistor circuit. The symbol for an inverter is shown in Figure 10.4, and is fairly self-explanatory. The triangle represents the fact that the circuit acts as what is called a buffer — that is, connecting the input to the output of a preceding stage does not affect or dampen the preceding stage in any way. The dot on the output tells us that an inverting action has taken place, and so the output state is the opposite of the input state in terms of logic level.

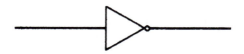

**Figure 10.4    The symbol of an inverter – the simplest form of digital logic gate**

To aid our understanding of digital logic circuits, input and output states are often drawn up in a table — known as a truth table, which maps the circuit's output states as its input states change. The truth table for an inverter is shown in Figure 10.5. As you'll see, when its input state is 1 its output state is 0. When its input is 0, its output is 1 — exactly as we've already hypothesised.

| IN | OUT |
|----|-----|
| 1  | 0   |
| 0  | 1   |

**Figure 10.5** Truth table of an inverter – the circuit of Figure 10.1, and the symbol of Figure 10.4

It's important to remember that, in both a logic symbol and in a truth table, we no longer need to know or care about voltages. All that concerns us is logic states — 0 or 1.

We can build up a simple circuit on breadboard to investigate the inverter and prove what we've just seen in theory. Figure 10.6 shows the circuit, while Figure 10.7 shows the circuit's breadboard layout. The digital integrated circuit used in Figure 10.6 and Figure 10.7 is actually a collection of six inverters, all accessible via the integrated circuit's pin connections — such an integrated circuit is commonly called a hex inverter, to signify this. We've simply used one of the integrated inverters in our circuit.

**Figure 10.6** Circuit of an experiment to discover how an inverter works

**Figure 10.7    Possible breadboard layout of the circuit in Figure 10.6**

The integrated circuit used is a type of integrated circuit called a CMOS device, and it's in what's known as the 4000 series of integrated circuits — it's actually known as a 4049. Don't worry about this for now, as we'll look at more devices in the series (and another series, for that matter) later in the chapter. All you need to know for now is that it's a standard dual-in-line (DIL) device, so you must be careful that you position it correctly in the breadboard.

Making sure that the actual integrated inverter used in our circuit has the correct input (that is, either logic 1 or logic 0) is easy. Logic 1 is effectively the same as a connection to the positive battery supply — in practice though, you should never connect an input 'directly' to the positive battery supply; instead always connect it using a resistor. On the other hand, logic 0 is just as effectively the same as a connection to the negative battery supply (this time though, it's perfectly allowable to connect an input directly). In other words, when the inverter's input is connected to positive it's at logic 1: when connected to negative it's at logic 0.

# Digital integrated circuits I

It's not a good idea to leave an input unconnected (what is called 'floating'), as it may not be at the logic level you expect it to be, so we've added a fairly high-valued resistor to the input of the inverter to connect it directly to the positive battery supply. This means that, under normal circumstances — that is, with no other input connected — we know that the inverter's input is held constantly at logic 1. To connect the inverter input to logic 0, it's a simple matter of connecting a short link between the input and the negative battery supply. This isn't shown in the breadboard layout, but all you need to do is add it when you are ready to perform the experiment.

Measuring the output of the inverter is just as easy, and we take advantage of the fact that the output can be used to light up a light-emitting diode (LED). So, when the inverter's output is at logic 1, the LED is lit: when the inverter's output is at logic 0, the LED is unlit.

Once the circuit's built, it's a simple matter of carrying out the experiment to note the output as its input changes, and completing its truth table. Figure 10.8 is a blank truth table for you to fill in. Just note down the circuit's output, whatever its input state is. Of course, your results should tally exactly with the truth table in Figure 10.5 — if it doesn't, you've gone wrong, because as I've said before in this book — I can't be wrong, can I?

| IN | OUT |
|----|-----|
|    |     |
|    |     |

Figure 10.8    Empty truth table for your results of the experiment in Figure 10.6

## Other logic circuits

The inverter is the first example of a type of digital circuit that's normally called a gate. Other logic gates are, in fact, based on the inverter and are made up from adaptations of it. There are a handful of logic gates that make up all digital circuits, from simple switches through to complex computers.

Other logic gates we need to know about are covered over the next few pages. They are all based on the principles of digital logic we've already looked at, though. However, before we discuss them, it's important for us to see that they're all based on a particular form of logic, which is essentially mathematical, known as Boolean algebra.

## Boolean algebra

Inputs to and outputs from digital circuits are denoted by letters of the alphabet A, B C and so on. Each input or output, as we have seen, can be in one of two states: 0 or 1. So, for example, we can say A = 1, B = 0, C = 1 and so on.

Now, of course, in any digital system, the output will be dependent on the input (or inputs). Let's take as an example the simplest logic gate — the inverter — we've already seen in operation. Figure 10.9 shows it again, along with its truth table, only this time the truth table uses standard Boolean algebra letters A and B. If A = 1, then we can see that B = 0. Conversely, if A = logic 0, then B = 1.

216

# Digital integrated circuits I

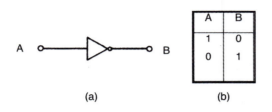

|   A   |   B   |
|:-----:|:-----:|
|   1   |   0   |
|   0   |   1   |

(a)          (b)

**Figure 10.9   (a) the symbol for an inverter or NOT gate; (b) its truth table**

In Boolean algebra we write this fact much more succinctly, as:

$$A = \overline{B}$$

where the bar above the letter B indicates what is called a NOT function — that is, the logic state of A is NOT the logic state of B. Alternatively, we could equally well say that B is NOT A, that is:

$$B = \overline{A}$$

Both Boolean statements are true with respect to an inverter, and are actually synonymous. Incidentally, inverters are often called NOT gates for this very reason.

You should be able to see that the Boolean statements for a gate simply allow us a convenient method of writing down the way a gate operates — without the necessity of drawing a gate symbol or forming the gate's truth table. No matter how complex a gate circuit is, if we reduce it to a Boolean statement, it becomes simple to understand.

217

## Other logic gates

The other logic gates we need to know about are every bit as simple as the NOT gate or the inverter. They all have a symbol, they all have a truth table, but most important, they all have a Boolean statement we can associate with them.

### OR gate

The symbol of an OR gate is shown in Figure 10.10. To see how an OR gate works, however, we'll conduct an experiment to complete a truth table. The circuit of the experiment is shown in Figure 10.11, while the breadboard layout is in Figure 10.12.

**Figure 10.10** **The circuit symbol for a 2-input OR gate**

**Figure 10.11** **Experimental circuit to find out how an OR gate works**

218

**Figure 10.12   Possible breadboard layout for the experimental circuit showing OR gate operation in Figure 10.11**

Build up the circuit, and fill in the blank truth table given in Figure 10.13 according to the results of the experiment as you change the circuit's inputs. You should see that the output of the circuit is logic 1 whenever one or other of the circuit's inputs are logic 1. Let's say that again — whenever one 'or' the other — which is why it's called an OR gate.

The results you should have got for your truth table are shown in the truth table of Figure 10.14.

| A | B | OUT |
|---|---|-----|
| 0 | 0 | |
| 0 | 1 | |
| 1 | 0 | |
| 1 | 1 | |

**Figure 10.13   A truth table to put the results of your experiment into**

| A | B | OUT |
|---|---|-----|
| 0 | 0 | 0 |
| 0 | 1 | 1 |
| 1 | 0 | 1 |
| 1 | 1 | 1 |

**Figure 10.14   The results you should have got in your experiment**

An OR gate is not quite as simple as an inverter — an inverter, after all, has only one input. An OR gate has more than one input — in theory it can have any number of inputs but, obviously, as the number of inputs increases, the resultant truth table gets more complicated.

Figure 10.15, shows a three input OR gate symbol and truth table. Being a three input gate, its truth table has eight variations. We can calculate this because, being binary digital, the circuit can have $2^3 = 8$ variations of inputs.

| A | B | C | Z |
|---|---|---|---|
| 1 | 1 | 1 | 1 |
| 1 | 1 | 0 | 1 |
| 1 | 0 | 1 | 1 |
| 1 | 0 | 0 | 1 |
| 0 | 1 | 1 | 1 |
| 0 | 1 | 0 | 1 |
| 0 | 0 | 1 | 1 |
| 0 | 0 | 0 | 0 |

(a)                                    (b)

**Figure 10.15   (a) the symbol for an OR gate; (b) its truth table**

220

# Digital integrated circuits I

The truth table of the OR gate shows that output Z of the gate is 1 when either A OR B OR C is 1. In Boolean terms, this is represented by the statement:

$$Z = A + B + C$$

where '+' indicates the OR function and has nothing whatsoever to do with the more usual arithmetical 'plus' sign. Don't be confused: just remember that every time you see the '+' sign in Boolean algebra it means OR.

## AND gate

As you might expect, an AND gate follows the Boolean AND statement. Figure 10.16 shows an AND gate circuit, while its breadboard layout is shown in Figure 10.17, and Figure 10.18 is a blank truth table ready for you to record the experiment's results. As you might already expect, an AND gate is so called because its output is at logic 1 when input 1 AND input 2 are at logic 1. As a result, your experimental truth table should look like the complete truth table shown in Figure 10.19.

**Figure 10.16  Circuit to investigate AND gate Boolean operation**

221

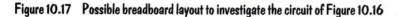

**Figure 10.17   Possible breadboard layout to investigate the circuit of Figure 10.16**

| A | B | OUT |
|---|---|-----|
| 0 | 0 | |
| 0 | 1 | |
| 1 | 0 | |
| 1 | 1 | |

**Figure 10.18   Truth table for you to record your results of your experiment**

| A | B | OUT |
|---|---|-----|
| 0 | 0 | 0 |
| 0 | 1 | 0 |
| 1 | 0 | 0 |
| 1 | 1 | 1 |

**Figure 10.19   How your results should look**

Translating this into a Boolean statement for a three input AND gate with inputs A, B, and C (shown as a symbol and as a truth table in Figure 10.20), we can see that its output Z is 1, only when inputs A AND B AND C are 1. The Boolean statement is therefore:

$$Z = A.B.C$$

where '.' indicates the Boolean AND expression.

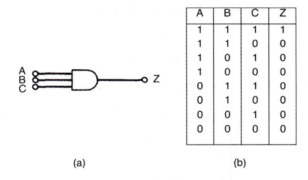

| A | B | C | Z |
|---|---|---|---|
| 1 | 1 | 1 | 1 |
| 1 | 1 | 0 | 0 |
| 1 | 0 | 1 | 0 |
| 1 | 0 | 0 | 0 |
| 0 | 1 | 1 | 0 |
| 0 | 1 | 0 | 0 |
| 0 | 0 | 1 | 0 |
| 0 | 0 | 0 | 0 |

(a)                         (b)

**Figure 10.20  (a) a three input AND gate; (b) its truth table**

## NOR gate

The NOR gate's function can be calculated from its name 'NOR'. Effectively, we can think of this as meaning 'NOT OR'. In other words, it's an OR gate and a NOT gate (that is, an inverter) in series.

The circuit of our experiment to look at NOR gates is shown in Figure 10.21, while a possible breadboard layout for the circuit is given in Figure 10.22. A blank truth table for you to fill in is shown in Figure 10.23.

**Figure 10.21   Circuit to find out the function of a NOR gate**

**Figure 10.22   Possible breadboard layout to build the circuit of Figure 10.21**

| A | B | OUT |
|---|---|-----|
| 0 | 0 |     |
| 0 | 1 |     |
| 1 | 0 |     |
| 1 | 1 |     |

**Figure 10.23   Blank truth table to complete with the results of your experiment**

| A | B | OUT |
|---|---|-----|
| 0 | 0 | 1 |
| 0 | 1 | 0 |
| 1 | 0 | 0 |
| 1 | 1 | 0 |

**Figure 10.24  A completed truth table for the experiment in Figure 10.21**

In Boolean terms, the output of a NOR gate is the exact inverse of the OR gate's output. When applied to the three input NOR gate and its truth table shown in Figure 10.25, we can see that the NOR gate's output is 1 when neither A NOR B NOR C is 1 — in other words, the exact inverse of the OR gate's output.

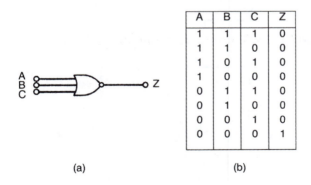

| A | B | C | Z |
|---|---|---|---|
| 1 | 1 | 1 | 0 |
| 1 | 1 | 0 | 0 |
| 1 | 0 | 1 | 0 |
| 1 | 0 | 0 | 0 |
| 0 | 1 | 1 | 0 |
| 0 | 1 | 0 | 0 |
| 0 | 0 | 1 | 0 |
| 0 | 0 | 0 | 1 |

(a)                              (b)

**Figure 10.25   (a) a NOR gate; (b) its truth table**

As a Boolean statement this is written:

$$Z = \overline{A + B + C}$$

# Starting electronics

### NAND gate

You should be able to work out for yourself what the NAND in NAND gate stands for: NOT AND. So, if you've followed the last bit about NOR gates, you might also be able to work out its Boolean statement. Nevertheless, it's best to see things for ourselves, and the circuit shown in Figure 10.26 is the experiment we need to carry out to understand what's happening. Figure 10.27 is a possible breadboard layout for the circuit in Figure 10.26.

**Figure 10.26  Circuit to investigate a NAND gate**

**Figure 10.27  Breadboard layout to build the circuit of Figure 10.26**

Meanwhile Figure 10.28 is a blank truth table for you to fill in with your results.

| A | B | OUT |
|---|---|---|
| 0 | 0 | |
| 0 | 1 | |
| 1 | 0 | |
| 1 | 1 | |

**Figure 10.28  Blank truth table, ready for the results of your experiment**

Figure 10.29 is the completed truth table for a NAND gate, and yours should be the same (if not, why not, eh?).

| A | B | OUT |
|---|---|---|
| 0 | 0 | 1 |
| 0 | 1 | 1 |
| 1 | 0 | 1 |
| 1 | 1 | 0 |

**Figure 10.29  Completed truth table for a NAND gate**

And, for the sake of completeness, Figure 10.30 shows a three input NAND gate and its truth table.

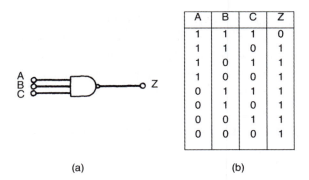

| A | B | C | Z |
|---|---|---|---|
| 1 | 1 | 1 | 0 |
| 1 | 1 | 0 | 1 |
| 1 | 0 | 1 | 1 |
| 1 | 0 | 0 | 1 |
| 0 | 1 | 1 | 1 |
| 0 | 1 | 0 | 1 |
| 0 | 0 | 1 | 1 |
| 0 | 0 | 0 | 1 |

(a)                              (b)

**Figure 10.30  A three input NAND gate (a) symbol, (b) truth table**

Actually, it's quite difficult to express a NAND gate's operation in words, but here goes: it's a gate whose output Z is 1 when NOT A AND NOT B AND NOT C is 1. Err... yeah, right!

A better way is to express it as the Boolean statement:

$$Z = \overline{A.B.C}$$

which, no doubt you have already worked out!

Haven't you?

## Simple, eh?

Earlier on I mentioned that all digital circuits can be made up from the simplest of digital circuits — the inverter. I'm going to prove that now; theoretically at first, then we can do some experiments which show the fact practically.

By expanding the inverter circuit we first looked at back in Figure 10.1 and the following circuits, we can easily create other circuits. Figure 10.31, for example, is merely a transistor operating in the same way that the inverter transistor circuit of Figure 10.1 works.

It is, effectively, an inverter with three inputs to its base, rather than just one.

228

**Figure 10.31   A transistor switch, with three inputs A, B and C**

If you look at it carefully, you should be able to figure out how it works. When all three of the three inputs, A, B, or C are connected to logic 0, the transistor is off, hence the transistor is a very high resistance, which means that the output is at logic 1. However, when any one of the three inputs is connected to logic 1 the transistor turns on, hence becoming a low resistance, and the output becomes logic 0. The truth table of Figure 10.32 shows this, and if you compare this to the truth table in Figure 10.25 you'll see they are the same. Yes, the three input transistor switch of Figure 10.31 (which is merely an adapted inverter) is a NOR gate!

| A | B | C | OUTPUT |
|---|---|---|--------|
| 1 | 1 | 1 | 0 |
| 1 | 1 | 0 | 0 |
| 1 | 0 | 1 | 0 |
| 1 | 0 | 0 | 0 |
| 0 | 1 | 1 | 0 |
| 0 | 1 | 0 | 0 |
| 0 | 0 | 1 | 0 |
| 0 | 0 | 0 | 1 |

**Figure 10.32   Truth table for the transistor switch circuit of Figure 10.31**

229

# That ol' black magic!

To show that the inverter is the heart of all other logic gates, we have to consider how other gates are made. We've just seen that we can make a NOR gate from an inverter, but what about the other logic gates?

## NOR and NOT gates combined

Figure 10.33 is a circuit that combines two logic gates we've already experimented on. They're linked so that the output of one (a NOR gate) can be used as the input of the next (an inverter or NOT gate) to produce a single device.

**Figure 10.33 A simple circuit to create one type of gate (OR) from two other types (NOR and NOT)**

A possible breadboard layout is shown in Figure 10.34, while an incomplete truth table is shown in Figure 10.35. Note that, as there are two stages to the circuit — that is, two separate logic gates — we can measure the middle section of the circuit by including another LED to more fully understand what's going on inside the whole thing. Because of this, of course, the truth table needs to have an extra column to allow us to record the actions of the circuit as we carry out the experiment. Figure 10.36 shows a complete truth table for the circuit, which your results should mirror.

**Figure 10.34 A breadboard layout for the experiment shown in Figure 10.33**

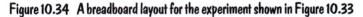

| A | B | C | Z |
|---|---|---|---|
| 1 | 1 |   |   |
| 1 | 0 |   |   |
| 0 | 1 |   |   |
| 0 | 0 |   |   |

**Figure 10.35 A truth table for you to record the results of your experiment**

| A | B | C | Z |
|---|---|---|---|
| 1 | 1 | 0 | 1 |
| 1 | 0 | 0 | 1 |
| 0 | 1 | 0 | 1 |
| 0 | 0 | 1 | 0 |

**Figure 10.36  A completed truth table for the experiment of Figure 10.33**

If we compare the truth tables (in either Figure 10.35 or Figure 10.36) with the truth table in Figure 10.14 we should see that the outputs are the same for the same inputs. (OK, yes, the inputs and output are presented in a different order — but the results are identical.)

In other words, we have used an inverter along with a NOR gate (by merely inverting the output of the NOR gate) to create an OR gate.

For completeness, Figure 10.37 shows a three input device based on this circuit, together with its truth table.

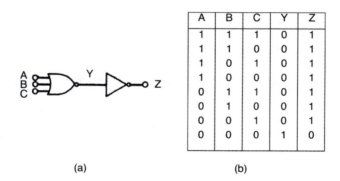

| A | B | C | Y | Z |
|---|---|---|---|---|
| 1 | 1 | 1 | 0 | 1 |
| 1 | 1 | 0 | 0 | 1 |
| 1 | 0 | 1 | 0 | 1 |
| 1 | 0 | 0 | 0 | 1 |
| 0 | 1 | 1 | 0 | 1 |
| 0 | 1 | 0 | 0 | 1 |
| 0 | 0 | 1 | 0 | 1 |
| 0 | 0 | 0 | 1 | 0 |

(a)  (b)

**Figure 10.37  (a) using an inverter and a NOR gate to create an OR gate; (b) its truth table to prove this**

## OR and NOT gates combined

In the same vein as the last circuit, Figure 10.38 shows a circuit combining different types of logic gates — this time an
OR gate whose inputs have been inverted by NOT gates. The
outputs of both NOT gates (hence, also, the inputs of the OR
gate) are measured with two LEDs, while the OR gate output
is measured with the third LED.

**Figure 10.38  A circuit to investigate OR and NOT gates**

Figure 10.39 shows a breadboard for the circuit for you to
follow. Figure 10.40 shows an incomplete truth table for you
to record your results of the experiment, while Figure 10.41
is a complete truth table, which — if everything goes as it
should — your results will be identical to.

If you've already been thinking about this, you'll no doubt
already have been flicking back through this book's pages to
locate which truth table Figure 10.40 and 10.41 are identical
to. If not, or if you simply haven't found it, it's the truth table
in Figure 10.29 — which is that of a NAND gate.

# Starting electronics

**Figure 10.39  A possible breadboard layout for the circuit of Figure 10.38**

| A | B | C | Y | Z |
|---|---|---|---|---|
| 1 | 1 |   |   |   |
| 1 | 0 |   |   |   |
| 0 | 1 |   |   |   |
| 0 | 0 |   |   |   |

**Figure 10.40  An incomplete truth table to record your experimental results**

| A | B | C | Y | Z |
|---|---|---|---|---|
| 1 | 1 | 0 | 0 | 1 |
| 1 | 0 | 0 | 1 | 1 |
| 0 | 1 | 1 | 0 | 1 |
| 0 | 0 | 1 | 1 | 0 |

**Figure 10.41  Complete truth table for the circuit of Figure 10.38**

# Digital integrated circuits I

For the sake of completeness, Figure 10.42a shows a three input variant of this, and the accompanying truth table in Figure 10.42b shows the various logic levels in the circuit.

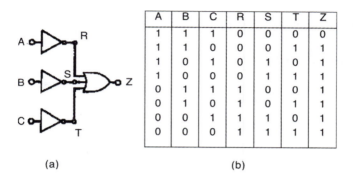

| A | B | C | R | S | T | Z |
|---|---|---|---|---|---|---|
| 1 | 1 | 1 | 0 | 0 | 0 | 0 |
| 1 | 1 | 0 | 0 | 0 | 1 | 1 |
| 1 | 0 | 1 | 0 | 1 | 0 | 1 |
| 1 | 0 | 0 | 0 | 1 | 1 | 1 |
| 0 | 1 | 1 | 1 | 0 | 0 | 1 |
| 0 | 1 | 0 | 1 | 0 | 1 | 1 |
| 0 | 0 | 1 | 1 | 1 | 0 | 1 |
| 0 | 0 | 0 | 1 | 1 | 1 | 1 |

(a)                                 (b)

**Figure 10.42  (a) a NAND gate, made from OR and NOT gates; (b) its truth table, to prove this**

## The Boolean Way — It's Logical!

Now we already know that the Boolean statement for a NAND gate is:

$$Z = \overline{A.B.C}$$

But we can also derive another Boolean statement from the above circuit, by the very fact that the inputs to the OR gate in Figure 10.42 are inverted before they reach the OR gate, such that the final output Z is equal to logic 1 when NOT A OR NOT B OR NOT C are logic 1.

# Starting electronics

So, the Boolean statement:

$$Z = \overline{A} + \overline{B} + \overline{C}$$

also expresses the NAND gate.

In other words:

$$\overline{A} + \overline{B} + \overline{C} = \overline{A.B.C}$$

Interesting!

## NOR from AND and NOT

And, just for the sake of completion, we can also work out that a NOR gate can be created by inverting the inputs of an AND gate.

As we've done so far in this book, we'll do this first experimentally, building up the circuit and checking results. Then we'll do it mathematically.

Figure 10.43 shows a circuit we can use to do this. It's basically a two input AND gate, the inputs of which have been inverted by NOT gates.

Figure 10.44 shows a possible breadboard layout you can follow to build the experimental circuit on, while Figure 10.45 is an uncompleted truth table for you to complete with your results, and Figure 10.46 is the complete truth table whose results should match yours.

**Figure 10.43   The experimental circuit to prove that a NOR gate can be created from an AND gate whose inputs are inverted first**

**Figure 10.44   A breadboard layout for the circuit in Figure 10.43**

| A | B | C | Y | Z |
|---|---|---|---|---|
| 1 | 1 | | | |
| 1 | 0 | | | |
| 0 | 1 | | | |
| 0 | 0 | | | |

**Figure 10.45 An incomplete truth table for you to record your results**

| A | B | C | Y | Z |
|---|---|---|---|---|
| 1 | 1 | 0 | 0 | 0 |
| 1 | 0 | 0 | 1 | 0 |
| 0 | 1 | 1 | 0 | 0 |
| 0 | 0 | 1 | 1 | 1 |

**Figure 10.46 A completed truth table for the circuit in Figure 10.43 – showing how we have made a NOR gate from an AND gate with NOT gates at its inputs**

And, for the sake of completeness, Figure 10.47 shows a three input AND gate with NOT gates at its inputs, and the circuit's truth table, to show that we can make a NOR gate of any number if inputs from an AND gate of that number of inputs together with the requisite number of inverters.

On the other hand, of course, and even better than this, by the same means of deduction we used for other circuits combining logic gates, we can calculate that the standard NOR Boolean statement:

$$Z = \overline{A + B + C}$$

is also the same as another Boolean statement:

$$\overline{A}.\overline{B}.\overline{C}$$

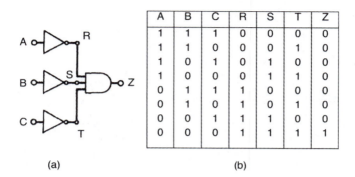

| A | B | C | R | S | T | Z |
|---|---|---|---|---|---|---|
| 1 | 1 | 1 | 0 | 0 | 0 | 0 |
| 1 | 1 | 0 | 0 | 0 | 1 | 0 |
| 1 | 0 | 1 | 0 | 1 | 0 | 0 |
| 1 | 0 | 0 | 0 | 1 | 1 | 0 |
| 0 | 1 | 1 | 1 | 0 | 0 | 0 |
| 0 | 1 | 0 | 1 | 0 | 1 | 0 |
| 0 | 0 | 1 | 1 | 1 | 0 | 0 |
| 0 | 0 | 0 | 1 | 1 | 1 | 1 |

(a)                           (b)

**Figure 10.47  A three input AND gate converted to a NOR gate with NOT gates**

In other words:

$$\overline{A}.\overline{B}.\overline{C} = \overline{A + B + C}$$

And — there — we have just proved mathematically that you can make a NOR gate from AND and NOT gates.

So, there's no doubt about it, all digital electronic logic gates can be created from a number of other digital electronic logic gates.

What's more, it's not beyond the realms of possibility to work out for ourselves that these very logic gates, that can make any other logic gates, can be combined, and combined again, to make up ever more complex circuits. And, indeed they are — as we'll see in the next chapter! For now, have a go at the quiz over the page, to see if you've been taking notice...

...which I'm sure you have. Just prove it!

# Starting electronics

## Quiz

Answers at end of book

1  In the circuit symbol for an NOT gate:

   a  the input is applied to the side with a circle
   b  the output comes from the side with a circle
   c  the circle implies that the signal is turned upside down
   d  the triangle represents the three signals that must be applied
   e  b and c
   f  all of these
   g  none of these.

2  All logic gates can be made up from transistors:

   True or false.

3  The output of a NOT gate is:

   a  always logic 0
   b  always 5 volts
   c  never logic 1
   d  the logical opposite of the power supply
   e  the logical opposite of its input
   f  none of these.

4  The output of an OR gate is logic 1 when:

   a  all of its inputs are logic 1
   b  any of its inputs are logic 1
   c  none of its inputs are logic 1
   d  a and b.

5  The output of a NOR gate is logic 1 when:

   a  all of its inputs are logic 1
   b  10 milliamps
   c  any of its inputs are logic 1
   d  this is a trick question, the output is always logic 1
   e  none of these.

6  When the output of a gate is used as the input of a NOT gate, the final output is:

   a  always the inverse of the output of the first gate
   b  always the same as the output of the first gate
   c  can only be calculated mathematically by binary logic
   d  all of these
   e  none of these
   f  in the lap of the gods.

7  If the inputs to an OR gate are inverted with NOT gates, the overall gate is that of a:

   a  NOT gate
   b  NOR gate
   c  NAND gate
   d  a and b
   e  b and c
   f  none of these.

8  Any logic gate can be created from other logic gates:

   True or false.

# 11   Digital integrated circuits II

In the last chapter we saw how electronic logic gates can be constructed from other electronic logic gates, however, as far as the hobbyist or the person starting out in electronics is concerned, knowing that logic gates can be created this way is one thing. No-one in their right mind would actually go about creating anything other than the simplest of logic gates like that, because it's just too time-consuming and too expensive. ICs represent the only sensible option.

Now we're going to take a look at some IC logic gates, in the form of the same sort of ICs we've already studied and experimented with. The ICs shown over the next few pages represent only the tip of the digital IC iceberg though, as I've only shown some of the most common and simplest. The digital circuits we'll look at later in this chapter are all available in one IC form or another as well, so the list of available ICs is very long indeed — not just the few shown here.

Starting electronics

## IC series

For most people, there are two families of ICs that are used in digital electronics. The first is a very easy-to-handle variety (in other words, they can't be easily damaged). On the other hand, they require a stable operating voltage of 5 V, which isn't all that easy to generate, without a power supply.

The second family of digital ICs will operate over a range of voltages (typically, a 9 V battery does the job exceedingly well), but are a little more easily damaged. Nevertheless, handled carefully and correctly, both families work well.

### 7400 series

The 7400 series of digital ICs is the variety that requires a fixed 5 V power supply. While there are many, many devices in the family only a few are of interest here. Before I explain about the ICs in detail though, a brief word about the 7400 series numbering.

The first two digits of any device in the family (that is, 74–) indicate that the device is a member of the family. The 7400 series is a family of digital ICs known as transistor–transistor logic (TTL) devices. The last two (or sometimes three) digits (–00) indicate which IC this is in the family. In other words a 7400 (which happens to be an IC with four, two input NAND gates) is different to a 7402 (which has four, two-input NOR gates).

A selection of ICs in the 7400 series is shown in Figure 11.1. Note that those shown — as I've already said — represent only a small fraction of the complete family.

Figure 11.1 Some integrated circuits from the 74 series of transistor–transistor logic (TTL) digital devices

## 4000 series

Like the 7400 series, the devices in the 4000 series are represented by their numbers. Note, though that the numbers between the two families do not correspond — in other words, a 4001 does not contain the same logic gates as a 7401.

# Starting electronics

A small part of the 4000 series is shown in Figure 11.2.

**Figure 11.2** **Some common devices from the 4000 series of digital integrated circuits — note that these cannot be directly interchanged with devices from the 74 series of TTL digital integrated circuits**

Once we have a family (well, two families actually) of digital integrated circuits that contain logic gates, it's fairly obvious that building more complex digital circuits becomes easier. It is merely a matter of using the gates within these integrated circuits to build more complex circuits. On the other hand, further devices in both ranges of integrated circuits are themselves more complex, containing circuits built from the gates themselves.

# Digital integrated circuits II

## Shutting the 'stable gate

All the digital electronic circuits we've looked at so far function more-or-less instantly. If certain inputs are present, a circuit will produce an output which is defined by the Boolean statement for the circuit — it doesn't matter how complex the circuit is, a combination of inputs will produce a certain output.

For this reason, such circuits are known as combinational. While they may be used in quite complex digital circuits they are really no more than electronic switches — the output of which depends on the correct combination of inputs.

Put another way, they display no intelligence of any kind, simply doing what is demanded of them — instantly.

One of the first aspects of intelligence — whether in the animal world (including humans) or in the electronics world — is memory (that is, the ability to decide a course of action dependent not only on applied inputs but also on a knowledge of what has previously taken place).

Digital circuits can easily be made that remember logic states. We say they can store binary information. The family of devices that do that in digital electronics are known as bistable circuits (called this, because they remain stable in either of two states), sometimes also known as latches. It's this ability to remain stable in either of the two binary states that allows them to form the basis of digital memory.

# Starting electronics

## SR-type NOR bistable

The simplest bistable or latch is the SR-type (sometimes called RS-type) bistable, the operation of which acts like two gates connected together as in Figure 11.3. The circuit is constructed so that the two gates are cross-coupled — the output of the first is connected to the input of the second, and the output of the second is connected to the input of the first.

**Figure 11.3    An SR-type bistable circuit, comprising two cross-coupled NOR gates**

There are two inputs to the circuit, and also two outputs. The inputs are labelled S and R (hence, the reason why the circuits are called SR-type...), which stand for 'Set' and 'Reset'. Now, it so happens that one output is, in most instances, the inverse of the other, so for convenience one output (by convention) is labelled Q and the other output (also by convention) is labelled $\overline{Q}$ (called bar-Q), to show this.

Both inputs to the SR-type NOR gate bistable should normally be at logic 0. As one input (either S or R) changes to logic 1, the output of that gate goes to logic 0. As the gates are cross-coupled, this is fed back to the second input of the other

246

gate, which makes the output of the other gate go to logic 1. This, in turn, is fed back to the second input of the first gate, forcing its output to remain at logic 0 — even (and this is the important point) after the original input is removed. This is an important point!

Applying another logic 1 input to the first gate has no further effect on the circuit — it remains stable, or latched, in this state.

On the other hand, a logic 1 input applied to the other gate's input causes the same set of circumstances to occur but in the other direction, causing the circuit to become stable, or latch, in the other way.

This sort of switching backwards and forwards in two alternate but stable states accounts for yet another term that is often applied to bistable circuits — they are commonly called 'flip-flops'.

The SR-type NOR bistable operation can be summarised in a truth table, in a similar way that non-bistable or combinational logic gates can be summarised. However, because we are trying to tabulate inputs and outputs that change with time, then strictly speaking we should call this a 'function table', and such a function table for the SR-type NOR bistable is shown in Figure 11.4.

Account has been taken in the function table for changing from one logic state to the other, by the symbol ⌐, which

indicates that the input state changes from a logic 0 to logic
1. Also, we no longer write the logic states as 1 and 0, but H
(meaning high) and L (meaning low).

| S | R | Q | $\overline{Q}$ |
|---|---|---|---|
| ⌐ | L | H | L |
| L | ⌐ | H | H |
| L | L | $Q_0$ | $\overline{Q}_0$ |
| H | H | L | L |

**Figure 11.4    The function table of the SR-type NOR bistable circuit shown in
Figure 11.3**

The outputs indicate what happens as the input states change
on these occasions. You can see that if input S goes high (that
is, from 0 to 1) while input R is low, then the output Q will be
high. If, however, input S then returns low, the output will be $Q_0$
— which simply means that it remains at what it was. If input
R then goes high while S remains low, the output changes to
low and remains in that state even if R goes low again.

The last line of the function table is shaded to indicate that
this condition is best avoided — simply because both outputs
are the same. Designers of digital circuits would normally take
steps to ensure that this condition did not occur in their logic
circuits because confusion in the form of uncertain outcomes
can obviously arise.

# Digital integrated circuits II

In other words (and, yes, I know that it was an awful lot of words), we have constructed a form of electronic memory, one of the outputs of which can be set to a high logic state by application of a positive-going pulse to one of its inputs. It's a small, but, very significant advance — changing combinational logic gates into bistable circuits that can be used together to create sequences of logic level variations. In effect, what we have now done is to cross over from combinational logic circuits to a new type of logic circuits known as sequential logic circuits, with this simple but highly significant and very important addition of memory. Whatever we want to call them: bistables; latches; or flip-flops, they are sequential, and they can be used to create even more complex logic circuits.

## SR-type NAND bistable

In Figure 11.3 the gates are both NOR gates, but they can be just as simply NAND gates, as shown in Figure 11.4. This gives us an SR-type bistable made from NAND gates. Easy enough — but there are some important considerations.

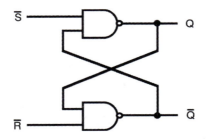

Figure 11.5    An SR-type bistable circuit comprising two cross-coupled NAND gates – note the inverted inputs when compared with Figure 11.3

The most important consideration is that both inputs of the SR-type NAND bistable should normally be at logic 1 — this is in direct contrast with the SR-type NOR bistable, where the inputs needed to be normally at logic 0. Because of this, the inputs are considered to be inverted in the circuit, shown as such in Figure 11.5.

With both inputs of the circuit of Figure 11.5 at logic 1, the circuit is stable, with both NAND gate outputs at logic 0. When one input goes to logic 1, the output of that NAND gate goes to logic 1. This is applied to the other NAND gate's second input, so the output of the second NAND gate will go to logic 0. This output is in turn fed to the first NAND gate's second input and thus its output is forced to remain at logic 1.

Applying a further logic 0 to the first NAND gate has no further effect on the circuit — it remains stable. However, when a logic 0 is applied to the second NAND gate's input causes the same reaction in the other direction. So the circuit has two stable states, hence is a bistable.

Figure 11.6 shows the function table for this SR-type NAND bistable.

Just as in the SR-type NOR bistable, there is a possibility that both outputs can be forced to be the same level which produces an indeterminate outcome. In the SR-type NAND bistable, this is when both inputs are at logic 0. So, circuit designers must ensure this situation does not occur in their logic circuits.

| $\bar{S}$ | $\bar{R}$ | Q | $\bar{Q}$ |
|---|---|---|---|
| ⌐L | H | H | L |
| H | ⌐L | L | H |
| H | H | $Q_0$ | $\bar{Q}_0$ |
| L | L | H | H |

**Figure 11.6**    A function table for an SR-type NAND bistable circuit, such as that in Figure 11.5

While these simple gate circuits are really all that's required to build bistables in theory, practical concerns require that something is added to them before they work properly and as expected. The problem is that the very act of applying changing logic levels to its inputs suffers from the mechanical limitations of the connecting method. Any mechanical switch (even one you might use in your home to turn on the lights) has 'contact bounce' — a phenomenon where the contacts in a switch flex while the switchover takes place, causing them to make and break the connection several times before they settle down.

Now, in the home, this is unnoticeable, because it happens so quickly you couldn't see the light flickering due to it, even if the light itself could react that quickly. But digital circuits are quick enough to notice contact bounce, and as each bounce is counted as a pulse the circuit could have reacted many, many times to just one flick of the switch. While we were merely changing input states and watching output states of combinational logic circuits in the last chapter, this really didn't matter. All we were wanting was a fixed output for a set of fixed inputs. In sequential circuits like bistables — and the other sequential circuits we'll be looking at in this chapter, as well as others — contact bounce is a real problem, as we are wanting a sequence of inputs. The wrong sequence — given by contact bounces — will give us the wrong results. Simple as that!

The likes of simple bistables like the SR-type NAND bistable give us an ideal and effective solution to the problems created by contact bounce. We've just discussed how once one of the NAND gates in the circuit is switched by application of one logic 0 pulse at either it's S or R input, it then doesn't matter if any further pulses are applied to the input. So, a switch that provides the pulse could make as many contact bounces as it likes, the circuit will not be affected further.

Figure 11.7 shows the simple addition of a switch and two resistors to the SR-type NAND bistable, to give a circuit that can be used in other circuits to provide specific logic pulses that have absolutely no contact bounce.

Figure 11.7    Simple circuit to prevent contact bounce

# Digital integrated circuits II

Switch SW1 in the circuit of Figure 11.7 is a simple single-pole, double-throw (SPDT) switch, and should have break-before-make contacts (in other words, when switching from one connection to the other, its switch contacts will disconnect from one connection before it connects to the other — so there is a point in the switch motion when all three contacts are disconnected from each other). The switch can be either a push-button switch, or a conventional flick-type switch.

Operation is fairly simple. At rest, one of the NAND gate inputs connects through the switch to logic 0, while the other NAND gate's input is connected through a resistor to logic 1. As the switch is switched, the second NAND gate's input is connected to logic 0 though the switch, and the circuit operates as previously described like a standard SR-type NAND bistable. It doesn't matter whether there is contact bounce or not at this time, as repeated pulses do not affect the bistable.

When the switch is operated again, so the first NAND gate's input is connected to logic 0 and the bistable jumps to its other stable state. Again, contact bounce will make no difference to it.

One or both of the outputs of the simple de-bouncing circuit can be applied to the inputs of following circuitry, safe in the knowledge that contact bounce will have no effect. The requirement for no contact bounce is true for any sequential circuit, actually. So this very simple solution can be used not just in the simple circuits we see here in this chapter, but for all sequential circuits. Whenever a sequential logic circuit requires a manually pulsed input, this circuit can be used.

## Other bistables

There is more than one type of bistable — the SR-type is in reality only the simplest. They all derive from the SR-type bistable however, and so it's the SR-type bistable that we'll use as the basic building block when making them.

### The clocked SR-type bistable

Figure 11.8 shows the clocked SR-type bistable. It is a simple adaptation of the basic SR-type NAND bistable, in that it features an extra pair of NAND gates, which not only allows inputs which do not require inverting, but also allows a clocked input to be part of the circuit's controlling mechanism.

Now, a clock circuit can be made from something like a 555 astable multivibrator (as we saw in Chapter 5), and how it is made is unimportant to us here. What is important, on the other hand, is that a clock means that several bistables can be synchronised, all in time with the clock.

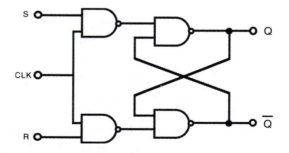

**Figure 11.8    A clocked SR-type NAND bistable**

# Digital integrated circuits II

The clocked SR-type NAND bistable circuit here functions basically as before, except that the outputs can only change states while the clock input CLK is at logic 1. When the clock input CLK is at logic 0, it doesn't matter what logic levels are applied to the S and R inputs, there will be no effect on the bistable.

Thus, with this simple addition, we have created a bistable with a controlling input. Unless the controlling input is at logic 1, nothing else can happen. This can be a very useful thing in electronics where, say, a number of things need to be counted over a period of time. For example, by clocking the circuit for, say, 1 hour, the number of cars passing a checkpoint in that time could be counted electronically.

## The D-type bistable

A variation on the basic clocked SR-type bistable is the 'data' bistable, or D-type bistable. A circuit is shown in Figure 11.9.

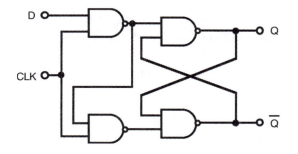

**Figure 11.9    A D-type bistable**

255

Effectively, the output of the first NAND gate (which would be the inverted S input) is used as the input to the second (which would be the R input). The single remaining input is given the label D.

In operation, the Q output of the D-type bistable will always follow the logic level at D while the clock signal CLK is at logic 1. When the clock falls to logic 0, however, the output Q remains at the last state of the D input prior to the clock changing.

This is quite a useful circuit, and is relatively simple in terms of both circuitry and operation. Like the simple SR-type bistable, it can be used to measure input signal variations over a defined period of time.

## The edge-triggered SR-type bistable

However, often it's necessary that things are registered at a precise instant of time, not over a period. So the circuits of Figures 11.8 and 11.9 can't be used. A moderately simple adaptation to ensure a circuit only registers input changes at an instant is all that's required — it's basically a combination of two identical clocked SR-type NAND bistables, each operating on opposite halves of the clock signal — and the resultant circuit is shown in Figure 11.10.

Such a combination of two bistables is often called a 'master–slave' bistable, because the input bistable operates as the master section, while the output bistable is slaved to the master during half of each clock cycle.

**Figure 11.10  An edge-triggered SR-type NAND bistable**

An important component is the the inverter which connects the two bistables in the circuit. This ensures that the bistables are enabled during opposite half cycles of the clock signal.

Assuming that the clock input CLK is at logic 0 initially, the S and R inputs cannot yet affect the master bistable's operation. However, when the clock input CLK goes to logic 1 the S and R inputs are now able to control the master bistable in the same way they do in Figure 11.8. As the inverter has inverted the clock signal though, the slave bistable's inputs (which are formed by the outputs of the master bistable) now have no effect on the slave's outputs. In short, although the outputs of the master bistable may have changed, they do not yet have any effect on the slave bistable.

When the clock input CLK falls back to logic 0 the master bistable once again is no longer controlled by its S and R inputs. At the same time, however, the inverted clock signal now allows the slave bistable's inputs to control the slave bistable. In other words, the final outputs of the circuits can only change state as the clock signal CLK falls from logic 1 to logic 0. This change of state from logic 1 to logic 0 is commonly called the 'falling edge', and the overall circuit is generically known as an 'edge-triggered' bistable.

This is an extremely important point in electronic terms. By creating this master–slave bistable arrangement to make the bistable edge-triggered, we are able to control precisely when the bistable changes state. As a benefit, this also makes sure that there is plenty of time for the master and slave bistables comprising the overall bistable to respond to the input signals — although things in logic circuits change and respond quickly, they do not happen instantly and still do take a finite time. The master–slave arrangement takes account of and caters for this small but finite time.

### The JK-type bistable

One other problem which we've already encountered with our basic bistables isn't yet catered for though — the indeterminate output which can occur in a bistable if both S and R inputs are logic 1 at the moment when the clock signal falls from logic 1 to logic 0.

So, to prevent this happening, it's a matter of preventing both S and R inputs from being at logic 1 at the same time as the clock signal falls from logic 1 to logic 0. We do this by adding some feedback from the slave bistable to the master bistable, and creating new inputs (labelled J and K).

The circuit of such a JK-type bistable to perform this function is shown in Figure 11.11.

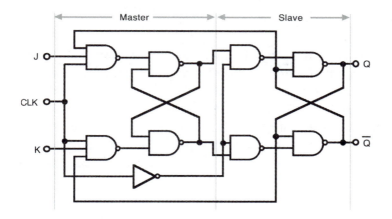

**Figure 11.11    A JK-type bistable**

As with the edge-triggered master–slave SR-type bistable, the outputs only change on the falling edge of the clock CLK signal, so the inputs (J and K now, not S and R) control the output states at that time. However, the feedback from the final output stage back to the input stage ensures that one of the two inputs is always disabled — so the master bistable cannot change state back and forth while the clock input is at logic 1. Instead, the enabled input can only change the master bistable state once, after which no further change of states can occur.

Because the JK-type bistable is completely predictable in this manner, under all circuit conditions, the JK-type bistable is the preferred minimum bistable device for logic circuit designers. That's not to say that SR-type bistables can't be used, and in fact they do have their purposes, but the important point is that circuit designers have to be aware of their limitations, ensuring that unpredictable outcomes are not allowed and so are designed out of the circuit.

## T-type bistable

One particular mode of operation of the JK-type bistable is of importance here. If both J and K inputs are held at logic 1, the outputs of the master bistable will change state with each rising edge of the clock signal and the final outputs will change state for each falling edge. Such a fact is used as the basis of the T-type bistable (where T stands for 'toggle'), shown in Figure 11.12. The lone T input of this circuit is really just the CLK inputs of other bistables.

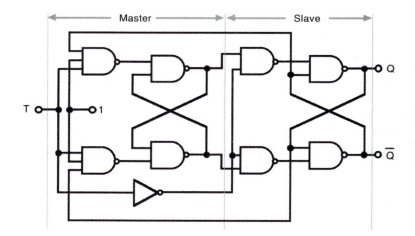

**Figure 11.12    A T-type bistable**

And that concludes our look at how logic bistable circuits are made up from basic logic gates. The important point about all of this though, is that by taking basic combinational logic gates and joining them in what are fairly basic circuits, we have created sequential circuits — that is, logic circuits that have an element of memory!

260

# Digital integrated circuits II

## Open the black box

Just as we found earlier that we could create logic gates with collections of transistors, but decided we could draw logic gates with their own symbols (rather than actually drawing all the internal transistors involved), so there are symbols for the various types of bistable — that don't show the actual logic gates (or, indeed, the individual transistors) involved.

Figure 11.13, for example, shows the symbol for a clocked, edge-triggered, SR-type bistable, such as the circuit of Figure 11.10. It's just a box, with the bistable's main inputs shown on the box. Note though, that it doesn't show whether the SR-type bistable is created from NAND gates or NOR gates (or whatever gates, or indeed whatever transistors, actually — that's all very irrelevant!). All you need to remember is that it's just a box showing inputs and outputs — and that it acts as a clocked, edge-triggered, SR-type bistable.

**Figure 11.13    A clocked, edge-triggered, SR-type bistable symbol**

Of note is the symbol used for the clock input, >, which merely represents the fact that it is edge-triggered. Its function table is shown in Figure 11.14.

261

| S | R | Q | $\overline{Q}$ |
|---|---|---|---|
| ⌐ | L | H | L |
| L | ⌐ | H | H |
| L | L | $Q_0$ | $\overline{Q}_0$ |
| H | H | L | L |

**Figure 11.14   The function table of the clocked, edge-triggered, SR-type bistable of Figure 11.13**

Such an approach of simplifying circuits is very common in electronics, and we've come across it before — the black-box approach. The approach relies on the fact that we don't need to know how a complex circuit is constructed — we only need to know its function.

Every single one of the other logic gate types we've seen in this chapter can also be represented in black box block form by symbols, and to prove it, we'll consider them all now. All you have to do is remember that if necessary they can — and sometimes are — built using logic gates, but you don't need to know how to construct them that way because — yes, you've guessed it — they can be bought ready-built in IC form, in the same series (7400 and 4000) we've already used and are familar with.

First is the D-type bistable. Its circuit symbol is shown in Figure 11.15, while its corresponding function table is in Figure 11.16.

**Figure 11.15  A D-type bistable symbol**

| $\overline{PR}$ | $\overline{CL}$ | CK | D | Q | $\overline{Q}$ |
|---|---|---|---|---|---|
| L | H | X | X | H | L |
| H | L | X | X | L | H |
| L | L | X | X | H | H |
| H | H | ⌐ | H | H | L |
| H | H | ⌐ | L | L | H |
| H | H | L | X | $Q_0$ | $\overline{Q_0}$ |

**Figure 11.16  The function table of a D-type bistable, such as the one shown in Figure 11.15**

Note that the inputs $\overline{PR}$ (preset) and $\overline{CL}$ (clear), are included in this D-type bistable (unlike the D-type bistable shown in Figure 11.9). These perform similar functions to the S and R inputs of the SR-type bistable. You should now be able to understand the function table. The first three rows indicate that the $\overline{PR}$ and $\overline{CL}$ inputs override all other inputs (that's why the clock and D inputs are shown as X: because X = don't care). Thus the outputs Q and $\overline{Q}$ are dependent primarily upon the logic states of the $\overline{PR}$ and $\overline{CL}$ inputs.

# Starting electronics

However, when the $\overline{PR}$ and $\overline{CL}$ inputs are both high, control of output becomes the prerogative of the clock and D inputs. For example, when the clock is low, the output $Q = Q_0$. That is, it remains the same as it was — no matter what the logic state of the D input (as $D = X$). On the leading edge of the clock pulse, however, output Q follows the logic state of the input D; whether high or low.

## JK-type bistable symbol

The JK-type bistable symbol is shown in Figure 11.17, while its function table is shown in Figure 11.18.

**Figure 11.17    A JK-type bistable**

| $\overline{PR}$ | $\overline{CL}$ | CK | J | J | Q | $\overline{Q}$ |
|---|---|---|---|---|---|---|
| L | H | X | X | X | H | L |
| H | L | X | X | X | L | H |
| L | L | X | X | X | H | H |
| H | H | ⎍ | L | L | $Q_0$ | $\overline{Q}_0$ |
| H | H | ⎍ | H | L | H | L |
| H | H | ⎍ | L | H | L | H |
| H | H | ⎍ | H | H | TOGGLE |

**Figure 11.18    The function table of a J-K master–slave bistable, such as that shown in Figure 11.17**

264

There's a couple of points to note about this JK-type bistable symbol:

- operations take place on the falling edge of the applied clock pulse;

- the last row of the function table shows a bistable operation known as toggling — in which the output state changes from whatever state it is, to the other state. This, you will of course remember — is how the T-type bistable of Figure 11.12 is created.

## And now the time has come...

So with this (fairly detailed) look at digital integrated circuits and the electronic logic that they comprise, thus ends our foray into the enjoyable world of electronics. You may feel as though you've learned a lot in the pages of this book and I trust you've had a good time in the process, but hopefully this is just the beginning of enjoyable times ahead.

Electronics is a much larger area than you've seen here, and there's much more to discover and much more to do. Indeed, you could spend literally years reading about electronics, and designing or building projects, and you still mightn't know it all but at least you are on the right track now — you're starting electronics...

... though just in case you think you *already* know it all — check out the quiz over the page!

# Starting electronics

## Quiz

Answers at end of book

1   A bistable can be made from:

   a  two NOR gates
   b  two NAND gates
   c  an IC
   e  a and b
   f  b and c
   g  all of these
   h  none of these.

2   All bistables can be made up from transistors:

   True or false.

3   A function table:

   a  shows contact bounce in a circuit
   b  is like a truth table for sequential circuits
   c  never applies to bistables made from logic gates
   d  only applies to IC bistables
   e  a and b
   f  none of these.

4   Contact bounce:

   a  is a myth
   b  is impossible to eliminate practically
   c  always occurs on Saturday
   d  is none of these.

5   All clocked bistables are edge-triggered:

   True or false.

6   One way of making an edge-triggered bistable:

   a  is to combine two bistables in a master–slave arrangement
   b  is to adjust the clock signal so that it only has edges
   c  is to disconnect the battery
   d  all of these
   e  none of these

7   The T in T-type bistable stands for:

   a  Total
   b  Typical
   c  Transient
   d  Toggle
   e  Two sugars and milk
   f  none of these.

8   The JK-type bistable is useful as:

   a  it needs no power supply
   b  all possible outputs can be predetermined logically
   c  a doorstop
   d  it allows fewer inputs than an SR-type bistable
   e  a and c
   f  b and c
   g  none of these.

266

# Glossary

**180° out-of-phase**   when two identical sinewaves are upside-down with respect to each other.

**active**   a device which controls current is said to be active.

**ampere**   standard unit to measure electrical current. Abbreviated to amp. Symbolised by A.

**amplitude**   the amplitude of an a.c. signal is the difference in voltage between its middle or common point and the peak voltage.

**analogue**   one of the two operating modes of a transistor, in which a variable tiny base current is used to control a large collector current.

267

**astable multivibrator**   an oscillator whose output is a squarewave.

**avalanche point**   the point on a diode's characteristic curve when the curve rapidly changes from a low reverse current to a high one.

**avalanche breakdown**   the electronic breakdown of a diode when reverse biased to its breakdown voltage.

**avalanche voltage**   breakdown voltage.

**axial**   component body shape in which connecting leads come from each of the two ends of a tubular body.

**base current**   shortened form for base-to-emitter current of a transistor.

**base, emitter, collector**   the three terminals of a standard transistor.

**bias**   to apply a fixed d.c. base current to a transistor, which forces the transistor to operate partially, even when no input signal is applied.

**bias current**   the d.c. current applied to a transistor to force the transistor into partial operation at all times.

**bias resistor**   in a common emitter circuit, the resistor which is connected to the base of the transistor, through which the bias current flows.

**block**   to prevent d.c. signals passing through while allowing a.c. signal to pass — normally done with a capacitor.

**bode plots**   method of showing a circuit's frequency response, where straight lines are used to approximate the actual response.

**breadboard block**   tool which allows you to build circuits temporarily, and test them. When you have completed your tests and are sure the circuit is working, you can remove the components and re-use them.

**breakdown**   effect in a diode, when reverse biased, where the reverse voltage remains more or less constant with different reverse currents.

**breakdown voltage**   the reverse voltage which causes a diode to electronically break down.

**bridge rectifier**   a collection of four diodes, either discrete or within an IC, which is used to provide full-wave rectification of an a.c. voltage input.

**buffer**   (1)   a device with high resistance input and low resistance output, which does not therefore load a preceding circuit, and is not loaded by a following circuit; (2)   another term for voltage follower.

**can**   nickname for the metal body of a transistor.

**capacitor**   an electronic component consisting of two plates separated by an insulating layer, capable of storing electric charge.

**characterisitic equation**   the mathematical equation which defines the characteristic curve of an electronic component.

# Starting electronics

**circuit diagram**   method of illustrating a circuit using symbols, so that all electrical connections are shown but not physical ones.

**closed-loop gain**   the gain of an operational amplifier with feedback.

**collector current**   shortened form for collector-to-emitter current of a transistor.

**conventional current**   current which is assumed to flow from a positive potential to a more negative potential. Conventional current is, in fact, made up of a flow of electrons (which are negative) from a negative to more positive potential.

**corner frequency**   the frequency at which a signal size changes from one slope to another, when viewed as a graph of size against frequency. In the simple filter circuits in this article, the corner frequency, f, is given by the expression:

$$f \ = \ \frac{1}{RC}$$

**coulomb**   standard unit to measure quantity of electricity.

**current**   flow rate of electricity. Current is measured in amps.

**current gain**   $h_{fe}$, shortened forms for forward current transfer ratio, common emitter, which is the ratio:

$$\frac{\text{collector current}}{\text{base current}}$$

of a transistor.

# Glossary

**decibels**   logarithmic units of gain ratio.

**dielectric**   correct term for the layer of insulating material between the two plates of a capacitor.

**digital**   one of the two operating modes of a transistor, in which presence or lack of base current to the transistor is used to turn on or off the transistor's collector current.

**diode**   a semiconductor electronic component with two electrodes: an anode and a cathode, which in essence allows current flow in only one direction (apart from when break-down has occurred).

**diode characteristic curve**   a graph of voltage across the current through a diode.

**discrete**   term implying a circuit built up from individual components.

**electrolytic capacitor**   a capacitor whose function is due to an electrolytic process. It is therefore polarised and must be inserted into circuit the correct way round.

**exponential curve**   capacitor charging and discharging curves are examples of exponential curves. An exponential curve rises/falls to about 0.63/0.37 of the total value in one time constant, and is within 1% of the final value after five time constants.

**farad (F)**   the unit of capacitance.

**feedback**   when all or part of an op-amp's output is fed back to its input, to control the gain of the circuit.

**filter**   a circuit which allows signal of certain frequencies to pass through unaltered, while preventing passage of other signal frequencies.

**forward biased**   a diode is forward biased when its anode is at a more positive potential than its cathode.

**frequency**   the number of cycles of a periodic signal in a given time. Usually frequency is measured in cycles per second, or Hertz (shortened to Hz).

**frequency response curves**   graphs of a circuit's gain against the frequency of the applied a.c. signal.

**full-wave rectification**   rectification of an a.c. voltage to d.c., where both half-waves of the a.c. wave are rectified.

**half-wave rectification**   rectification of an a.c. voltage to d.c., where only one half of the a.c. wave is rectified.

**hertz**   the usual term for cycles per second.

**high-pass filter**   a circuit which allows signals of frequencies higher than the corner frequency to pass through unaltered, while preventing signals of frequencies lower than this from passing.

**hybrid transfer characteristic**   a graph of collector current against base-to-emitter voltage (for a transistor in common emitter mode).

**input characteristic**   a graph of base current against base-to-emitter voltage (for a transistor in common emitter mode).

# Glossary

**integrated circuit**   a semiconductor device which contains a chip, comprising many transistors in a complex circuit.

**inverting amplifier**   an op-amp with feedback, connected so that its input voltage is inverted and amplified.

**law of parallel resistors**   $\dfrac{1}{R_{ov}} = \dfrac{1}{R1} + \dfrac{1}{R2} + \dfrac{1}{R3} + \dots$

**law of series-resistors**   $R_{ov} = R1 + R2 + R3 + \dots$

**load**   a circuit which has a relatively low resistance input is said to load a preceding circuit, by drawing too much current.

**load line**   a line drawn on the same graph as a semiconductor's characteristic curve, representing the load's characteristic curve.

**low-pass filter**   a circuit which allows the passage of signals with frequencies lower than the corner frequency, but prevents the passage of signals with frequencies higher than this.

**multi-meter**   a test meter which enables measurement of many things e.g., voltage, current, resistance.

**non-inverting amplifier**   an op-amp with feedback, connected so that its input voltage is amplified.

**offset null terminals**   terminals on an op-amp which may be used to reduce or eliminate the offset voltage present at the op-amp's output.

**offset voltage**   the voltage present at an op-amp's output preventing it from equalling 0 V when the input is 0 V.

**ohmic** term used to refer to a component whose characteristic curve follows Ohm's law.

**Ohm's law** law which defines the relationship between voltage, current and resistance

**oscillator** a circuit which produces an output signal of a repetitive form.

**op-amp** abbreviation for operational amplifier.

**open-loop gain** an op-amp's gain without feedback.

**operating point** the point of intersection of a semiconductor's characteristic curve and the load line, representing the conditions within the circuit when operating.

**operational amplifier** a general-purpose amplifier (usually in integrated circuit form) which can be adapted and used in many circuits.

**output characteristic** a graph of collector current against collector-to-emitter voltage (for a transistor in common emitter mode).

**parallel** joined at both ends.

**passive** a device through which current flows or doesn't flow, but which cannot control the size of the current, is said to be passive.

**peak-to-peak voltage** the difference in voltage between the opposite peaks of an a.c. signal. Commonly shortened to p–p voltage, or pk–pk voltage.

**peak voltage**   the voltage measured at the peak of an a.c. signal.

**periodic**   any a.c. signal which repeats itself regularly over a period of time is said to be periodic.

**permittivity (e)**   a ratio of capacitance against material thickness, of a capacitor dielectric.

**phase shifted**   two similar a.c. signals which are out-of-phase i.e., start their respective cycles at different times are said to be phase shifted.

**pointer**   the indicator on a multi-meter.

**potentiometer**   a variable voltage divider used in electronic circuits. Many types exist.

**preset**   a potentiometer which is set on manufacture and not normally readjusted.

**quiescent current**   the standing current through a transistor due to the applied bias current.

**radial**   a component body shape in which both connecting leads come out of one end of a tubular body.

**range**   measurement which a multi-meter is set to read.

**reactance**   a physical property of a capacitor which may be likened to a.c. resistance.

**relative permittivity**   the ratio of how many times greater a material's permittivity is than that of air.

**relaxation oscillator**   an oscillator relying on the principle of a charging and discharging capacitor.

**resistance**   property of a substance to resist the flow of current. Measured in ohms.

**resistance converter**   another term for voltage follower.

**resistor**   electronic component used to control electricity. A large number of different types and values are available.

**reverse biased**   a diode is reverse biased when its cathode is at a more positive potential than its anode.

**ripple voltage**   the small a.c. voltage superimposed on the large d.c. voltage supplied by a power supply.

**saturation reverse current**   the small reverse current which occurs when a diode is reverse biased but not broken down.

**scale**   the numbers marked on a multi-meter which the pointer points to.

**self-bias**   a form of biasing a common emitter transistor which regulates any variance in the transistor's current gain.

**series**   joined end-to-end, in line.

**sinewave**   a constantly varying a.c. signal. Sinewaves are generally used to test electronic circuits as they comprise a periodic signal of one signal frequency.

**smoothing**   the process of averaging out a rectified d.c. wave, so that the resultant waveform is more nearly steady.

**squarewave**   a signal which oscillates between two fixed voltages.

**stabilise**   to produce a fixed d.c. voltage from a smoothed one.

**three-rail power supply**   a power supply which has a positive supply rail, a negative supply rail, and a 0 V supply rail.

**time constant (t)**   the product of the capacitance and the resistance in a capacitor/ resistor charging or discharging circuit.

**track**   the fixed resistive part of a potentiometer.

**transfer characteristic**   a graph of collector current against base current (for a transistor in common emitter mode).

**transistor**   a three layer semiconductor device, which may consist of a thin P-type layer sandwiched between two layers of N-type (as in the NPN transistor), or a thin layer of N-type material between two layers of P-type (as in the PNP transistor).

**transition voltage**   the knee or sharp corner of a diode characteristic when forward biased.

**volt**   standard unit to measure electrical potential difference. Symbolised by V.

**voltage**   potential difference — the flow pressure of electricity.

# Starting electronics

**voltage divider, potential divider**  a number of electronic components in a network (usually two series resistors) which allows a reduction in voltage according to the voltage divider rule.

**voltage divider rule**

$$V_{out} = \frac{R2}{R1 + R2} \times V_{in}$$

**voltage follower**  an op-amp circuit, in which the op-amp is connected as a unity gain, non-inverting amplifier.

**voltage regulator**  an integrated circuit which contains a stabilising circuit.

**wavelength**  the distance between two identical and consecutive points of a periodic waveform.

**wiper**  that part of a potentiometer which is adjustable along the resistive track.

**zener diode**  a special type of diode which exploits the breakdown effect of a diode when reverse biased.

**zero adjust knob**  the control on an ohmmeter which allows the user to zero the multi-meter.

**zeroing a multi-meter**  the adjustment made to an ohmmeter to take into account differences in voltage of the internal cell.

# Components used

## Resistors

2 x 150Ω

2 x 1k5

1 x 4k7

3 x 10 kΩ

2 x 15 kΩ

1 x 22 kΩ

3 x 47 kΩ

2 x 100 kΩ

1 x 220 kΩ

1 x 1k0 miniature horizontal preset

1 x 47 kΩ miniature horizontal preset

## Capacitors

1 x 1 nF

2 x 10 nF

2 x 100 nF

1 x 1 μF electrolytic

1 x 10 μF electrolytic

1 x 22 μF electrolytic

1 x 220 μF electrolytic

1 x 470 μF electrolytic

## Semiconductors

1 x 1N4001 diode

1 x OA47 diode

1 x 3V0 zener diode

3 x LED (any colour or type)

1 x 2N3053 transistor

1 x 555 integrated circuit timer

1 x 741 op-amp

1 x 4001 quad, 2-input NOR gate

1 x 4011 quad, 2-input NAND gate

1 x 4049 hex inverter

1 x 4071 quad, 2-input OR gate

1 x 4081 quad, 2-input AND gate

## Miscellaneous

battery (9 V — PP3-sized, or similar)

breadboard

multi-meter

single-pole single-throw switch (any type)

single-strand tinned copper wire (for use with breadboard)

# Starting electronics

## Quiz answers

| | |
|---|---|
| Chapter 1: | 1 d: 2 a; 3 e; 4 e; 5 c; 6 c |
| Chapter 2: | 1 c; 2 a; 3 d; 4 true; 5 e |
| Chapter 3: | 1 d; 2 e; 3 b; 4 true; 5 true; 6 a |
| Chapter 4: | 1 a; 2 f; 3 c; 4 c; 5 c |
| Chapter 5: | 1 a; 2 e; 3 c; 4 c |
| Chapter 6: | no quiz |
| Chapter 7: | 1 d; 2 f; 3 c; 4 c; 5 true; 6 b |
| Chapter 8: | 1 g; 2 a; 3 e; 4 b; 5 d |
| Chapter 9: | 1 a; 2 f; 3 d; 4 e; 5 true; 6 false |
| Chapter 10: | 1 b; 2 True; 3 e; 4 d; 5 e; 6 a; 7 c; 8 True |
| Chapter 11: | 1 g; 2 True; 3 b; 4 d; 5 False; 6 a; 7 d; 8 b |

# Index